"*Water, Creativity and Meaning* makes an insightful contribution to current under-standings of human–environmental relationships. Centering on creative practices, it explores the intimate and interconnected engagements with water that people experience and embody at a personal and local level, showing how these generate important memories and meanings; enable the composition of individual and community identities; and encourage deep and affective relations with place."

— *From the Foreword, Veronica Strang, University of Durham, UK*

"Beyond the empirical richness of the collection the most striking themes for me that emerge from the essays are a range of conceptual explorations at the leading edge of water research. I was very interested to see emerging interest in different forms of "attunement", attentiveness, and materiality, posing questions in terms of research methodology as well as the interpretation of different kinds of developments that span human and other-than-human realms."

— *From the Afterword, Matthew Gandy, Professor of Cultural and Historical Geography, University of Cambridge, UK*

Water, Creativity and Meaning

At a time of great turmoil and crisis, environmentally, socially and politically, water has emerged as a topic of huge global concern. Moreover, many argue that what is needed in order to change our relationship with the environment is a cultural paradigm shift. To this end, this volume brings together diverse approaches to exploring human relationships with the watery world and the other living things that rely upon it.

Through exploring multiple creative ways of engaging with water and people, the volume adds to the current zeitgeist of writing about water by expanding the discussion about this vital substance and how, as humans, we relate to it. Chapters focus on creative explorations and explorations of creativity in relation to developing these understandings, including concepts such as hydrocitizenship and responses to drought and flooding. Drawing on the in-depth research and experience of arts practitioners including participatory artists, as well as academics from a variety of fields including geography, anthropology, health studies and environmental humanities, the book provides a rich and multidisciplinary perspective on water and creative ways of engaging and understanding human–water relationships.

It represents a valuable source and inspiration for academics, arts practitioners and those involved in environmental policy and governance.

Liz Roberts is a cultural geographer currently working at the University of West of England, UK, on an RCUK project on digital storytelling and water scarcity (www.dryproject.co.uk).

Katherine Phillips is a social researcher at the University of West of England, UK, with interests in sustainable living, human–environment relations and the mainstreaming of radical alternatives.

Water Management, Food Security and Sustainable Agriculture in Developing Economies
Edited by M. Dinesh Kumar, M.V.K. Sivamohan and Nitin Bassi

Freshwater Ecosystems in Protected Areas
Conservation and Management
Edited by Max C. Finlayson, Jamie Pittock and Angela Arthington

Participation for Effective Environmental Governance
Evidence from European Water Framework Directive Implementation
Edited by Elisa Kochskämper, Edward Challies, Nicolas W. Jager and Jens Newig

China's International Transboundary Rivers
Politics, Security and Diplomacy of Shared Water Resources
Lei Xie and Jia Shaofeng

Urban Water Sustainability
Constructing Infrastructure for Cities and Nature
Sarah Bell

The Biopolitics of Water
Governance, Scarcity and Populations
Sofie Hellberg

Water, Technology and the Nation-State
Edited by Filippo Menga and Erik Swyngedouw

Revitalizing Urban Waterway Communities
Streams of Environmental Justice
Richard Smardon, Sharon Moran and April Baptiste

Water, Creativity and Meaning
Multidisciplinary Understandings of Human–Water Relationships
Edited by Liz Roberts and Katherine Phillips

For more information and to view forthcoming titles in this series, please visit the Routledge website: www.routledge.com/books/series/ECWRM/

Water, Creativity and Meaning

Multidisciplinary Understandings of Human–Water Relationships

Edited by Liz Roberts and Katherine Phillips

Routledge
Taylor & Francis Group
LONDON AND NEW YORK

from Routledge

First published 2019
by Routledge
2 Park Square, Milton Park, Abingdon, Oxon OX14 4RN

and by Routledge
52 Vanderbilt Avenue, New York, NY 10017

First issued in paperback 2020

Routledge is an imprint of the Taylor & Francis Group, an informa business

British Library Cataloguing-in-Publication Data
A catalogue record for this book is available from the British Library

Library of Congress Cataloging-in-Publication Data
Names: Roberts, Liz (Elisabeth), editor. | Phillips, Katherine, 1980- editor.
Title: Water, creativity and meaning : multidisciplinary understandings of human-water relationships / edited by Liz Roberts and Katherine Phillips.
Description: Abingdon, Oxon ; New York, NY : Routledge, 2018. |
Series: Earthscan studies in water resource management |
Includes bibliographical references and index.
Identifiers: LCCN 2018005653| ISBN 9781138087668 (hardback) | ISBN 9781315110356 (ebook)
Subjects: LCSH: Water–Social aspects. | Hydrologic cycle. | Water supply. | Water security. | Nature–Effect of human beings on.
Classification: LCC GB665 .W287 2018 | DDC 304.2–dc23
LC record available at https://lccn.loc.gov/2018005653

ISBN 13: 978-0-367-58788-8 (pbk)
ISBN 13: 978-1-138-08766-8 (hbk)

Typeset in Bembo
by Swales & Willis Ltd, Exeter, Devon, UK

Contents

Illustrations

Figures

Tables

Editor biographies

Liz Roberts is a cultural geographer currently working at the University of West of England on an RCUK project on digital storytelling and water scarcity (www.dryproject.co.uk). She undertook her PhD at the University of Exeter working on an AHRC-funded project titled 'Spectral Geographies of the Visual'. Liz's ongoing work has a focus on visual images and visual methods, the interaction of narrative and the visual, non-representational theory, cultural theory and philosophical and political engagements with the visual within the social sciences.

Katherine Phillips is a social researcher at the University of West of England, UK, with interests in sustainable living, human–environment relations and the mainstreaming of radical alternatives. Her PhD (Aberystwyth) explored governance, knowledge and aesthetics within an eco-village and low impact development movement in Wales. Most recently she has been researching human–water relationships through the AHRC project entitled *Towards Hydrocitizenship* (hydorcitizenship.com), using creative methods and participatory approaches to explore the wateriness of people and communities (including the more-than-human) in Bristol.

Contributor biographies

Jon Anderson is a Professor of Human Geography at the School of Geography and Planning, Cardiff University, UK. His research interests focus on the relations between culture, place and identity, particularly the geographies, politics, and practices that emerge from these. His key publications include: *Understanding Cultural Geography: Places and Traces* (Routledge, 2010, 2015), *Water Worlds: Human Geographies of the Ocean* (edited with K. Peters, 2014), and *Page and Place: Ongoing Compositions of Plot* (2014). Jon is principal investigator on the Arts and Humanities Research Council project, 'A New Literary Geography: Establishing a Digital Literary Atlas for Wales and its Borderlands' (April 2016–July 2018), and co-investigator on the Economic and Social Research Council project, 'Locality, Community and Civil Society', as part of the WISERD Civil Societies work package (September 2016–September 2018).

Lyndsey Bakewell is a research assistant in the Department of English and Drama at Loughborough University, UK, where she recently finished a PhD in the re-examination of spectacle in the Restoration period. Since completing her PhD Lyndsey has joined the storytelling team at Loughborough where she is considering the multiple ways in which storytelling and narrative can be used as a framework to assist in the examination and management of topical issues. Most recently Lyndsey has been working on projects relating to the environment, health and well-being, social inclusion and education.

Matt Birkinshaw is a PhD candidate at the London School of Economics and Political Science working on urbanisation, infrastructure and governance in South Asia. His doctoral research uses reform initiatives from government and the private sector to understand the politics of urban water in Delhi. Matt has around ten years of research experience, most recently at LSE Cities, and has worked with a range of international and community development NGOs. You can find more information at mattbirkinshaw.wordpress.com.

Maria Louise Bønnelykke is currently a postdoctoral associate at the Department of Learning and Philosophy at Aalborg University. Her current research is affiliated with ATSIP (Australian Tropical Science and Innovation Precinct) at James Cook University in Townsville, Australia where she currently studies interactions between humans and coral on the Anthropocene landscape of the Great Barrier Reef. Her PhD was an ethnographic study of perceptions of environmental change faced by islanders and development practitioners in the Pacific island nation Kiribati. Maria Louise is an environmental anthropologist with an interdisciplinary profile, and she has collaborated with engineers, natural scientists and computer programmers in solving problems in the private and the public sector.

Alison L. Browne is a lecturer in human geography and a research fellow at the Sustainable Consumption Institute (SCI) at the University of Manchester. Her research is primarily focused on understanding the dynamics of everyday life underpinning large-scale sustainability transitions (e.g. water, energy and food). Using a range of critical social science theory and mixed methodologies she plays with ideas of how everyday practices that consume resources come to be disrupted, changed and governed. Being deeply embedded in partnership working with non-academic stakeholders from government organisations through to large businesses, Alison's research has had a range of impacts in the UK water sector.

Rebecca L. Farnum is an environmental peacebuilding researcher and educator at King's College London. During her doctoral studies, she partnered with activist organisations to explore fog-harvesting, environmental journalism, and scuba diving conservation. As part of her preoccupation with all things hydro, Becca currently convenes the London Water Research Group. She also works on a project building capacity for young people in the green economy in Norfolk and Norway, for a network of social change activists in the Middle East and North Africa, and as a widening participation tutor with The Brilliant Club. www.rebeccalfarnum.com.

Ronan Foley is a senior lecturer at the Department of Geography at Maynooth University, Ireland. He has written extensively in the broad area of therapeutic landscapes, including the 2010 monograph *Healing Waters: Therapeutic Landscapes in Historic and Contemporary Ireland*; co-edited with Thomas Kistemann a 2015 special issue of *Health and Place* on healthy blue space; and recent papers on the relationships between swimming, health and well-being. He is currently the principal investigator on an Irish Environmental Protection Agency project on green/blue infrastructures and health as well as an advisory partner on an ESRC project at the University of Exeter on Sensing Nature.

Matthew Gandy is Professor of Geography at the University of Cambridge. His publications include *Concrete and Clay: Reworking Nature in New York City* (2002), *The Fabric of Space: Water, Modernity, and the Urban Imagination* (2014), and *Moth* (2016), along with articles in *New Left Review*,

International Journal of Urban and Regional Research, Society and Space and many other journals. He is currently researching the interface between cultural and scientific aspects to urban bio-diversity.

Luci Gorell Barnes began her professional life in the world of physical theatre but migrated to the realm of visual arts. Her work revolves around themes of childhood, isolation and belonging, and she writes, and makes books, maps and animated films to explore these ideas. Her participatory practice is concerned with those who find themselves on the margins for one reason or another and she develops responsive processes that help people to think imaginatively with themselves and others. Her creative collaborations contribute to a range of disciplines that include academic research, family support, health services and education.

John Hartley is an interdisciplinary artist, researcher and arts professional. His artworks explore ways and shapes of change in the substrates of water, technology and imagination through projects and exhibitions realised in galleries and the public realm across the UK and Europe. He co-founded and co-directed the arts agency Difference Exchange, working with artists and academics across the UK, Europe and Asia and previously worked at the Arts Council England on visual art, interdisciplinary and art and ecology projects. In 2016 he was awarded a practice-based PhD by the University of the Arts London, after undertaking research at Falmouth University. Since 2016 he has worked at the University of Exeter, developing its arts and culture activity.

Claire Hoolohan is a researcher at the Tyndall Centre for Climate Change, University of Manchester. Her research is focussed on understanding resource intensive practices and identifying opportunities for intervention. Recently she has worked on research projects on reducing domestic water use (her PhD thesis 'Reframing Water Efficiency' and related papers are available online), managing energy in food supply chains (published in the journals *Sustainable Production and Consumption* and *Energy Research and Social Science*), and understanding transitions in the water–energy–food nexus (Stepping Up project), working with non-academic research partners to understand how sustainable social practices can be facilitated.

Loraine Leeson is a visual artist known for socially engaged work with East London communities. Her projects have won a Media Trust Inspiring Voices award and her Olympic Inspire Mark public artwork involving 300 children, *The Catch*, became a London 2012 Landmark while *Active Energy* received RegenSW's Arts and Green Energy award. Loraine is director of cSPACE, Chair of Arts for Labour and runs an MA in Art and Social Practice at Middlesex University. In 2017 she exhibited her 1970s photomontage work in support of health campaigns at the ICA and published her book *Art: Process: Change – Inside A Socially Situated Practice*, with Routledge.

Antonia Liguori has a background in history and computer science and has a long-lasting experience in working with cultural institutions, museums and archives in Italy. Before moving to the UK, from 2006 to 2012, she coordinated the Multimedia Department at BAICR Sistema Cultura to develop innovative methodologies and digital environments to enhance ways of experiencing cultural heritage. In the past five years she has been involved in a variety of research projects at transnational level, including a fellowship at the Smithsonian Institution, to explore how to apply the digital storytelling methodology in different contexts (both social and cultural).

Antony Lyons is an eco-social creative practitioner with a background in environmental/geo-sciences and landscape design. Lyons is concerned with relationships between ecological change and human activities. Areas of particular focus include coastal/river landscapes, deep-time (geological) perspectives, routes/journeys and intangible cultural heritage. His creative research methods rely on fieldwork and experimental remixing of archives, recordings and contemporary narratives – explored in the context of both 'slow' and 'intensive' artist residencies. Responding to places, their possibilities and contestations, his work is expressed through sculpture, film, sound and intermedia installation. Lyons has worked with a number of academic research projects as an advisor as well as operating as an independent creative practitioner.

Ruth Macdougall studied sculpture and environmental art at the Glasgow School of Art and has worked on residencies and commissions under the auspices of several UK and international organisation including UNESCO, Historic Scotland, the International Committee for the Red Cross and the Tyndall Centre for Climate Change Research to name a few. Her work encompasses performance, installation, boat building and community development for young people. Through her master's studies she became immersed in the world of water security and continues to work with a worldwide network of water specialists, exploring culture and the communication of water stewardship through art. www.ruthmacdougall.work.

Simon Meisch heads the DFG junior research group 'Ethics of Science in the Research for Sustainable Development' at the International Centre for Ethics in the Sciences and Humanities at the University of Tuebingen (Germany). After studying politics and modern German literature in Tuebingen and Edinburgh, he did his PhD in politics. For many years he has been working in the fields of application-oriented ethics and research for sustainable development. His main research interests are water ethics, water security and the role of the humanities within the research for sustainable development.

Kirsten Rudestam studies water policy and management, 'sense of place' with respect to resource management practices and the role of affect and emotion in environmental politics. She investigates the relationship between water policies and everyday behaviours; how ideas about local waters are formed, the conflicts they produce, and how they enable or disable parti-cular forms of participation. Her current research examines more closely collaborative and participatory water management practices as they interface with tribal nations and Indigenous communities. She received her PhD in Sociology from the University of California, Santa Cruz, and currently teaches at Whitman College in eastern Washington.

Rob St John is an artist and academic. His interdisciplinary practice often involves landscape writing, art–science collaboration, sound recording and film-making, and has been shown at Tate Modern, London, Edinburgh Sculpture Workshop and Stour Space, London, alongside numerous publica-tions and releases. Rob is a PhD researcher in cultural geography at the University of Glasgow, working on art–geography exchanges, Anthropocene landscapes, and György Kepes' interdisciplinary practices and publications at MIT in the 1960s. www.robstjohn.co.uk.

Lyndsey Stoodley is a PhD student in the School of Geography and Planning, Cardiff University. Her research critically examines the World Surfing Reserve network and its role in protecting and celebrating the cultures and places of surfing. After graduating from Aberystwyth University in 2010 with a BScEcon International Relations, Lyndsey spent the next four years living and working in Germany, Australia, and China before returning to her hometown of Cardiff in 2014 to complete an MSc in Sustainability, Planning and Environmental Policy. A keen surfer, Lyndsey enjoys riding small waves on big boards in exotic locations like Indonesia and Porthcawl.

Charlie Thompson has an MSc in Water Security and International Develop-ment from the University of East Anglia. He has worked as a water resources scientist for over ten years with the US Geological Survey and the United Nations, researching the impacts of environmental contaminants to water resources. Currently, Charlie operates an international development consul-tancy in the United States and Ethiopia where he focuses on global water security research.

Michael Wilson is Professor of Drama in the School of Arts, English and Drama at Loughborough University. His main research interests lie broadly within the field of popular and vernacular performance. He has published widely on storytelling and led numerous RCUK and EU-funded projects that explore the application of storytelling to a variety of social and policy contexts, especially around environmental policy, health and social justice. He has been a member of the Advisory Boards for the Digital Economy Programme (RCUK, led by EPSRC), Connected Communities (AHRC) and Digital Transformations (AHRC). He is also Chair of the Arts and Humanities Panel for the British Council's Newton Fund programme.

Foreword

It has been both a pleasure and an honour to respond to the editors' invitation to write a foreword for this book. *Water, Creativity and Meaning* makes an insightful contribution to current understandings of human–environmental relationships. Focusing on creative practices, it explores the intimate and interconnected engagements with water that people experience and embody at a personal and local level, showing how these generate important memories and meanings, enable the composition of individual and community identities and encourage deep and affective relations with place.

This attention to the complex interconnections that characterise such immediate engagements with water highlights, in contrast, the ecological and ethical disconnectedness of 'Big Water': the vast infrastructures and techno-managerial systems that are so instrumentally focused on directing water flows to meet human needs and interests that the costs of doing so are largely – and unsustainably – externalised to other species and environments.

The location of examples at a grassroots level throughout the book is critical, providing insights into the rich knowledge of community groups, and how their interactions bring together diverse narratives and experiences. The particular focus on phenomenological engagements with water and its materialities also illuminates the ways that sensory experiences and creative relational practices can support health and engender appreciation of the world's own agentive capacities. This underlines a well-established reality: that intimate and holistic engagement with places, and with water bodies, often serves to elicit affective concern for their well-being.

In this sense, the contributions imply that, even in the most intensely industrialised and urbanised societies, personal, experiential engagements with water provide some emotional and imaginative common ground with the deeply affective environmental relationships maintained by many indigenous and other long-term, place-based communities. It is therefore entirely logical that the social and ecological activism promoted by immediate interactions with water draws much inspiration from indigenous and feminist critiques of modernity, and of neo-liberal efforts to reframe water reductively as a 'cooperative resource'.

The book is therefore timely in offering a fresh perspective on one of today's most pressing challenges: how can societies effect a paradigmatic transformation that will de-anthropocentrise human–environmental relations and enable more collaborative and convivial engagement with the non-human world? How can societies be persuaded to prioritise long-term sustainability over short-term exigencies and primarily economic aims? In 2017 I was asked to assist a United Nations High Level Panel on Water in composing some new Principles for Water to underpin the UN's Sustainability Goals. The major challenge was to translate diverse cultural and spiritual values relating to water into terms that would traverse international boundaries sufficiently to persuade heads of state to give more priority to such values in forming national water policies and practices.

Bringing deeper cultural relationships with water into public and international discourses is a challenge not just for anthropologists, but across the disciplinary spectrum. In essence, we all struggle to communicate complex ideas and relationships upwards in scale in ways that do not strip them of meaning and value. This volume demonstrates that the creative arts have a vital role to play, not just as an illustrative adjunct to academic research, but in co-producing and communicating knowledge in ways that resist reduction and fragmentation. Interdisciplinary efforts to co-produce new, more sustainable ways of thinking with and about water must therefore include the arts as equal partners in this endeavour.

Veronica Strang
Professor, Department of Anthropology, and Executive Director,
Institute of Advanced Study, University of Durham, UK

Preface

The original idea for this edited collection emerged from a series of conference sessions at the Royal Geographical Society's Annual International Conference in Exeter in 2015. The theme of the conference, chaired by Sarah Whatmore, was the Anthropocene, and we put out a call for papers that invited contributions on how creativity and creative approaches could enable a deepened understanding of human–water relationships within the Anthropocene. The impetus for this interest arose out of our research projects, first an Arts and Humanities Research Council funded project with the title 'Towards Hydrocitizenship'. The project, led by Professor of Environmental Humanities, Owain Jones, focused on using an arts and humanities interdisciplinary approach to community-embedded work looking at uncovering and expanding the narratives that individuals connect to in relation to water. A second project, led by Professor of Environmental Management Lindsey McEwen focused on the use of narratives in relation to understanding drought in a UK context. As researchers working on these two projects we were interested in exploring the ways in which creative methods and approaches to engagement with people in different contexts could provide insight into the fluid social dynamics of human–water relationships. To this end we brought people together at the conference for an afternoon of sessions focusing on different aspects of water, creativity and creative approaches.

As researchers coming in at a juncture in the social sciences wherein creative approaches and the notion of creativity more generally are being increasingly recognised as a source of energy and rupture, we felt placed in a unique position of attempting ourselves to grapple with the interstices of creativity, research and writing, to understand creative processes and how they rub up against different disciplinary academic approaches in Research Council-funded projects in which creative practice is involved. The insights generated by our contributors were expressed to us in other ways by many other creative practitioners and researchers on our projects. The challenge for those in the social sciences, and beyond that in hard sciences and technical and policy fields, is to really allow for completely different ways of thinking and doing that are unbounded by disciplinary conventions, that involve making unusual leaps and connections, that are relational, embedded and deeply and unapologetically contextual. This is not to say that more scientific disciplines lack creative thinking, but it is to present alternatives to the attempt to generalise or universalise and extract or abstract knowledge from context. Focusing on water in particular has allowed the volume to give space to the fluidity, contingency and relationality of knowledge. Water's inherent mutability and omnipresence in all life renders it a particular kind of vessel for the imagination and thought. In this volume we seek not to treat water as an object to be understood, but rather to recognise that through engaging with water we can uncover relational processes that allow us to think differently about our selves and our relationship to other forms of matter.

Acknowledgements

We are grateful to the many people involved in our projects and the RGS sessions who shared their thoughts and experiences. On the Towards Hydro-citizenship project, as we all muddled through in often quite murky waters, the insights of the creative practitioners on the project: Antony Lyons, Iain Biggs, Loraine Leeson, Simon Read, and the arts academics Stephen Bottoms and Tom Payne, expanded our conversations about creative practice and process and how it might interweave with academic practice. The Arts and Humanities Research Council provided the funds for colour images in the volume. There are numerous other people, as well as places, waterways, images, poems, films, plants, animals, riffles, mudbanks, estuaries, bogs, frogs and others that also inspired us in the gathering together of this book.

Introduction

A new era for human–water relationships

Katherine Phillips and Liz Roberts

In December 2017 the last episode of the BBC series *Blue Planet II* aired on television, viewed by an estimated 13 million people. The legendary David Attenborough, who for decades has brought the spectacle of nature into homes in the UK and beyond, spent this last episode focusing on some of the consequences of human activity, including the devastating effects of plastic, light and noise pollution in the world's seas on the creatures (including humans) that rely on them for life and livelihoods. *Blue Planet* is but one of many examples of media that suggest a slow but sure permeation of awareness and concern for watery environments into cultural and political discourse. The effects of anthropogenic climate change, and the increasing likelihood that geological strata from this epoch will indicate that the greatest influence on the planet has been that of humans, has inspired the dubbing of this era the 'Anthropocene' (Crutzen 2006). How we might as a species move towards more mutually beneficial interactions with the rest of the species, materials and energy on the planet is the subject of much debate. Yet it seems clear that besides increasing knowledge and awareness, there are cultural, emotive, conceptual and imaginative aspects to our relationships to the environment, including the watery world, influencing our behaviour and trajectories.

We aim with this collection to focus in on one key aspect of our material world, the interconnectedness of humans and water. Set within the context of a book series focused on 'water management', the volume draws attention to the shifting sense of meaning that water takes on in different contexts, embodiments and relationships. Here we present a collection of very human ways of relating to water, in the words of people coming from different disciplines, including geography, health studies, history and philosophy, but crucially also creative practitioners, reflecting on the growing trans-disciplinarity of watery research. Through the collection of writings we see how embracing creative ways of thinking and doing can, on a number of levels, enable the development of different relationships (with water, with a sense of self, with a sense of community and belonging). These explorations add to wider conversations about water and our relationship to it, as well as to conversations about the Anthropocene, but in a way that is accessible to those outside of disciplines that address these in depth.

Although the implications of the Anthropocene are debated, the idea that a paradigm shift is needed to address the overlapping social, environmental and political crises has a much longer history. Rachel Carson, famed for initiating environmental action with her writing of *Silent Spring* also wrote poignantly and poetically about the watery environments of the planet (Carson 1951; Carson and Hubbell 1998/1955) urging deeper appreciation and care. In the 1970s, the emergent notion of 'deep ecology' positioned humans as merely one part of a complex and interlinked system (Naess 1973). At the same time, ecofeminism linked societal domination and exploitation with that of the environment. These creative and philosophical understandings of the world continue to echo in contemporary voices that unsettle human–environment disconnections, e.g. Morton (2017) and Haraway (2016). In contrast, a technoscientific approach that largely removes poetical and the ethical questioning has dominated research, policy and material interactions with water. The domination of what some have termed modernity's influence on human–water relationships (Illich 1985; Linton 2010; Gandy 2014) has come to the attention of scholars in recent years. Such work indicates a move towards thinking aesthetically, creatively and responsibly in a way that acknowledges our senses and embodiment along with our capacity for intellectual reasoning, and seeks not to privilege the latter over the former.

Distributed creativity, thought-experiments and 'everyday' tactics

In this collection, creativity and creative practice are presented in a range of ways, including performative practices, thought experiments and in/ through 'everyday' practices such as the acts of reading or swimming. As suggested by Hallam and Ingold (2007), this reflects a conception of creativity that veers away from individualistic and pedestal notions of 'art' towards understandings of creativity as distributed, as social practice, as materially and contextually generative and as emerging out of the mundane as well as the exceptional.

The last 50 years has witnessed a re-focussing on notions of 'social practice' and 'social sculpture' as proposed by Joseph Beuys (Tisdall 1974; Thistlewood 1995) within the arts and social science. Much of this relates to paying closer attention to everyday interactions that explore the rituals, meanings, value systems and symbolism that influence behaviour. Geographers working with artists have picked up on these trajectories and have incorporated this knowledge into fields such as the geohumanities (see, for example, Hawkins 2015; Hawkins et al. 2015), recognising that creativity is not simply seen in end products, but speaks to a much wider ontological framing of humans as part of sets of relations that are always in a state of creative becoming. This state of becoming is understood as generative and creative. As Glaveanu notes:

> Human action is intrinsically creative (Joas 1996) because, at each point in time, it is embedded within a horizon of possibility. This space of possibility, however, is also at all times, constrained (by our intentions, by physical affordances, by cultural norms).
>
> (Glaveanu 2015, 168)

While human activity in general is therefore creative, it is often through moments of disruption and novelty that the horizons of possibility come into focus.

Creative social practice speaks to a rethinking of the artist as 'isolated creative genius' to a 'practitioner engaged in the world with an ecological perspective' (Gablik 1991 cited in Guyotte 2014, 13). Often, a practitioner will engage or collaborate with communities and interest groups, developing work that is process-oriented and a form of knowledge-making in its own right. Creative practitioners may adopt a 'practice as research' approach (Barrett and Bolt 2014), embodying performative and other creative elements and methods as part of their research practice. There are resonances here with 'participatory action research' (Kindon et al. 2007), and in both cases it should be noted that the participants need not be human. The chapters of this volume explore creativity both from the perspective of creative practitioners and from academics seeking more creative ways of participating in, and engaging with, people and water, doing so through exploring interactions and interfaces in different ways.

Watery being-ness that exceeds techno-managerial speak

Water is essential to all life as we know it. As such, it seems a good place to start with an attempt to think about the porosity of human experience. And yet, to call something as diverse as water by this elemental name seems to belie the intricate inter-weavings of life and water throughout existence. Water as both an environment and a material can be viewed as a creative agent. Its mutability allows it to be mist, ice, steam, hot, cold, transparent and opaque, deep, shallow, dispersed or solid, moving or still. Water can therefore be considered as a material par excellence in terms of its transformative potential. Importantly, water is not external to humans, nor humans to water:

> For us humans, the flow and flush of waters sustain our own bodies, but also connect them to other bodies, to other worlds beyond our human selves. Indeed, bodies of water undo the idea that bodies are necessarily or only human. The bodies from which we siphon and into which we pour ourselves are certainly other human bodies (a kissable lover, a blood

transfused stranger, a nursing infant), but they are just as likely a sea, a cistern, an underground reservoir of once-was-rain.

(Neimanis, 2017, 2)

As Neimanis points out, when it comes to water and humans, we are undeniably part of a flow. She suggests that 'Our watery relations within (or more accurately: as) a more-than-human hydrocommons thus present a challenge to anthropocentrism, and the privileging of the human as the sole or primary site of embodiment' (ibid.). Yet how do we reconcile our intellectual apprehension of water and watery worlds, as something outside of us, with our own watery beingness? And how, when we are characterised as an enviromentally destructive species, do we begin to understand ourselves as part of this 'more-than-human hydrocommons'?

Water, in various forms, captures the imagination, something that is evident in a recent run of popular books that focus on the joys of interactions with water (and implicitly nature) for instance through wild swimming (Deakin 2000) sea travel and intellectual exploration of the substance (Jha 2015), or learning how to 'read water' through observation (Gooley 2016). Alongside this celebratory nature writing however, both popular (e.g. Pearce 2006) and academic writing in recent decades has been framed in a discourse of crisis, scarcity and security (Bakker 2012). Bakker charts the exponential increase, for example, in writing on 'water security', a framing that is diverse and includes issues of geopolitical security alongside those of drought and access to potable water and sanitation. Water is therefore seen as variously, a giver of joy, food, leisure and spiritual connection, but also potentially in its absence or excess, the taker of life and livelihood.

In recognition of this complexity, water research is increasingly called upon to recognise multiple disciplinary outlooks and perspectives, as well taking into consideration social and cultural contexts (e.g. Strang 2004, 2010; Krause and Strang 2016). However, despite the clarion calls for 'integration', 'interdisciplinarity' 'multidisciplinarity', 'more-than-human research', etc., the outputs of scientific/technical and policy-related studies rarely mention the personal, the embodied, the intimate, the poetic and the spiritual, or they judge these to be unimportant to techno-political strategising over a precious resource.

Dominant discourses of water

Linton (2010) argues that 'modern water' as a discursive construction has been highly effective at harnessing and controlling extensive portions of the hydrosocial cycle, through the construction of dams, the draining of wetlands, vast irrigation projects and the diversion of water into circulatory systems in cities. These modifications have altered the ecology of the planet, contributing to anthropogenic climate change. Modern water, in Linton's view, is an understanding of water simply as a 'resource', stripped of any cultural, or even political and ethical connotations.

> Even the term 'water management' implies a particular kind of hydro-social relation, one characterized by deference to a particular kind of water, stripped of its complex social relationships such that it may be managed by experts who are not necessarily directly involved in these relationships.
>
> (Linton 2010, 58)

In Linton's view, water is not a neutral and external substance or object, but something that is intextricably woven into social, economic and political relationalities. Water has become bound up with a discourse that presents technological management and control as universally desirable and an apolitical construct. The result is progressive exacerbation of social inequality and damage to ecological systems.

Even within the technocratic realm of water management with its attendant notions of modern water, however, there is increasing recognition of the need to consider and incorporate multiple perspectives, and both human and wider ecological interactions with water. In response, we now observe significant amounts of scholarship devoted to the topic of 'water governance', with particular attention being paid to multiple 'stakeholders' and forms of 'participatory' and 'adaptive' management. Integrated water resource management (IWRM) has emerged from global institutional partnerships that now seek ways of integrating the voices of a wider range of water users in decision-making. The power relationships and implications of these movements, however, are also open to critique. Do these shifts represent a sufficient radical shift of understanding to enable the kind of changes needed, towards a far more sustainable and socially and environmentally just watery world?

The movement towards water governance is, at least in part, motivated by water conflicts and other socio-political consequences arising from what is framed as a global water crisis (for example see World Economic Forum 2017). Since the 1980s, Peter Gleick and others have warned that along with climate change and an increased incidence of drought, the risk of violent conflict is increasing (Gleick 2014). Conflict (or the potential for it) is certainly apparent in cases of 'transboundary' waters, and this too is the subject of much scholarship, as evidenced by even a cursory glance at other titles in this series (e.g. Abukhater 2013; Hanazs 2017; Mirumachi 2015; Xie and Shaofeng 2017). On a global scale, concerns about the depletion of groundwater, pollution and overuse of surface water sources and the impacts of climate change have led to claims that 'water is the new oil', a statement that seems to emphasise the status of water as a nationally vital resource, and one to be fought over.

Besides conflict and geopolitical insecurity, water is enmeshed in the immediate and everyday politics of human and more-than-human life. Maria Kaika, Erik Swyngedouw, Matthew Gandy and Karen Bakker among others have brought attention to the power relationships embedded and carried by water in cities, including the issue of 'produced scarcity' (Kaika 2005; Mehta 2010), and the injustice of discourses of 'resilience' to extreme weather (e.g. Welsh 2014). Injustice and inequality are made evident time and again in the use, access and distribution of water. As Marc Reisner (2017, 12) pointed out in his observation of water in the American West, 'water flows uphill to money'. The capitalist impulse that drives modern water is insensitive to environmental complexities and driven by ever-increased 'production', providing increased supply to pre-empt increased demand. Obstacles to supply are seen to be overcome using technology, such that for instance, the depletion of groundwater is addressed through the desalination of seawater. Water is reduced to an abstraction, a resource, a commodity and this view has limited space for the social, cultural or environmental dimensions that cannot be incorporated into the capitalist model.

Expanded cultural imaginations of water

> Water is what we make it.
>
> (Linton 2010, 3)

If it's true that water is what we make of it, then the 'we' is important. To attempt to understand the 'we', at least in a human sense, authors such as Veronica Strang (2004) have taken an anthropological approach; for instance exploring communities and their relationships with water via landscape, culture, religion and spirituality, private lives, water governance, agriculture, home and garden. The significance of the meanings and values of water, and the types of power relations that are embedded in these, regain attention through such approaches:

> How are some social actors able to impose their definition of water on other social actors with different but equally legitimate definitions? In other words, how is power used in the service of one or another of cultural definitions of water?
>
> (Donahue and Johnston cited in Linton 2010, 58)

The ways in which humans give meaning to water is a political act with far-reaching implications. However, while social actors are able to impose meanings on water, water too can be viewed as 'generative and agentive co-constituent of relationships and meaning in society' (Krause and Strang 2016, 633). Thinking in this way requires moving beyond the boundaries of discipline and even particular theoretical perspectives.

There is inspiration to be found through philosophical approaches in terms of how we might look differently at human–water relationships. Insights can be gleaned for instance from literatures on materiality (e.g. Bennett 2010), Actor Network Theory (e.g. Callon 1986; Law 2009) and more-than-human geographies (e.g. Bastian et al. 2017). The radical connectivities, or 'kin', presented by Donna Harraway (2016) and Neimanis' (2017) 'posthuman feminism for the Anthropocene' in particular offer insights into how we can rethink (and re-feel) our interconnections within the material world. These approaches argue for plurality and fluidity in understanding our being and existing. Agency is distributed, found in relations between things, rather than purely a characteristic of humans.

Coming from a background of anthropology, Veronica Strang argues that the properties of water exist through relations, meaning that the material agency of water may provide affordances that are combined differently with culturally and contextually specific meanings (Alberti 2014; Davies 2014). Her aim is to show how water's material agency is a challenge to the dominating nature–culture dualism and its alienation of human kind from 'the other'. It advances a political positioning: that 'putting non-human agency at the fore is thus seen to be an ethical imperative, one that can be used to challenge forms of resource exploitation' (Alberti 2014, 160; Strang 2014). It is this type of ethical imperative that we take as the basis of gathering the creative interventions presented in the chapters, as a means of representing a collective working through of ideas relating to the self, human communities and connections and relationships with the more-than-human, even where authors may not mention any of the specific philosophies refered to above.

Alongside these shifts in academic thinking, which may in some cases be occuring in quite rarified intellectual environments, there are more widespread movements occuring that pose a challenge to the dominant modernist paradigm, drawing attention to alternative modes of living with, and respecting, the biosphere. In recent years, indigenous peoples and communities have gained international recognition for their resistance to domination and their defence of land and water, sometimes paying for this with their lives. Iconic indigenous campaigner Berta Cáceres argued strongly not only for consideration of indigenous people whose lives and livelihoods are interlinked with the river, but also on behalf of the river itself:

> When we started the fight for Río Blanco, I would go into the river and I could feel what the river was telling me. I knew it was going to be difficult. But I also knew we were going to triumph, because the river told me so.
>
> Caceres (2015)

The efforts of indigenous movements to change relationships with water and the wider environment involve a different paradigm to that of modern water.

Instead of economic benefit from the mining powered by the hydroelectric dam, the language of these movements is often one of sacredness, of livelihoods and life and of intimate connection to place and environment. While indigenous knowledge has long been in the lexicon of academia, it has often been incorporated into a paradigm that remains techo-scientific, or studied as an anthropological curiosity. The shift within some academic spheres towards listening more closely to indigenous knowledge is, in our view, part of the struggle towards re-learning how to be in/with the world in the way that the Actor Network theorists, those focused on 'vibrant matter' or 'object oriented ontology', hydrofeminism and other strands are all aiming for. It represents a more serious consideration of the way that social injustice and environmental injustice are linked.

Making the links between modes of knowledge and exploitation is reflected in the move within academic geography institutions to 'decolonise' geography. This was the theme of the 2017 Royal Geographical Society's Annual Conference and engendered conversations on the need, as well as the tensions involved in, decolonising language along with decolonising geography. Decolonising water, from this position, would involve embedding practice in inquiry, so that inquiry is not about generating knowledge, but about ethical and responsible action rooted in a respect for people, waters and place (MacGregor 2017). It is easy to see how this relates to critiques of 'modern water', exploitation and domination through colonialism and/or capitalism and also how indigenous knowledge as well as movements can challenge ways of thinking and being that reinforce and recreate ruptured and disconnected socio-natures (Castree and Braun 2001).

Enfolding renewed ways of thinking and being 'with water' into the sphere of academia is a challenge in need of creative formulations. Imaginative practices have long viewed water as symbolic; holding spiritual significances and emotional resonances. There are strong water imaginaries across different cultures, and the multiple sensory modes of perception in relation to water are often central to these. Water and water sounds have been found to soothe, and washing and bathing are cultural behaviours that serve either to connect people with each other or to provide solace, comfort and connection for individuals in private spaces, reflective of class and racial distinctions. Water can present different material and cultural imaginaries in different contexts that are important to understand. Matthew Gandy (2014, 2) highlights the need for interdisciplinarity in considering the complexities of the cultural and material significance of water and relations between modernity, nature and the urban imagination, suggesting that establised disciplines associated with water management (engineering) or urban water systems (geography) can be complemented 'by insights drawn from fields such as anthropology, architecture, art history, comparative literature, epidemiology, hydrology, sociology and science and technology studies'. Through this collection, we wish to make a stronger case for a transdisciplinary creativity in thinking and being, and we feel that it is in these zones that some of the breakthroughs for human–water relationships can germinate and be nurtured.

Key to this is allowing ourselves to think outside of boundaries, to think (and feel) relationally (Krause and Strang 2016, 633), as embodied watery beings on a watery planet. Through forefronting more fluid and creative practices, we seek to overcome the paradox faced by attempting to critique modernity through an essentially modern format of language, disciplinary boundaries and academic conventions. To that end, this collection brings together the voices of academics and creative practitioners whose diverse approaches and interactions with watery environments and the people entangled within them draw on personal, embodied, intimate and lived experiences and watery conversations in different contexts.

Introduction to the chapters

The book is organised into four sections, with a view to giving space to different aspects of creative engagements with water.

Part I Fluid processes: creative research with water

In the first section, the four chapters – in different yet overlapping ways – draw attention to the nature of creative practices and processes. We have chosen to begin the book with a section that is explicitly reflective about creative practices. The chapters offer insight into how creative practice can surface alternative relationships with water and how creative facilitation can create space for different types of knowledge to come into contact. All of the chapters engage with senses of relationality.

In Chapter 1, Loraine Leeson finds water in the form of the River Thames emerges as a focal point for a community group of elderly men who are approached through a project about technology and older people. Their links with the local landscape and their creative thinking about the power of water emerge through the project as a focus on the energy-production potential of the tidal river. The starting point here is not the river and water, but this comes about naturally through interactions with the group. Yet perhaps the most significant point that Leeson shares is that as a creative practitioner she takes on a role of enabling and encouraging the flow of creativity among the people she works with, making, developing and enabling linkages to occur and take on new forms.

Creativity then, like water, can be an enabler of flows, carrying energy, directing and redistributing it. Leeson draws attention to the long-term relationships built up between herself and the group, as well as other people who become involved at various points. Her attendance to the practical realities of interaction and engagement as a creative practitioner allows insight into how such processes can enable the expression of creativity in unusual, and sometimes ordinary and everyday ways, often overlooked both by research that parachutes in artists to produce work, as well as in arts-funded projects looking for iconic art-world products.

In Chapter 2, Luci Gorell Barnes weaves water into her prose, as its mysteries become part of her conceptualisation and feelings about place, people and belonging. In this intimate and poetic chapter, Luci writes from the perspective of a wall, a stream and flood waters. This way of experimenting with points of view suggests attachments that extend beyond the human, intimating that places, other creatures (e.g. foxes) and even objects (e.g. a wall) may feel attachment to or a mutuality with the humans who interact with them. This creativity in both perception and expression allows a different lens on relationships with water and place that draws the reader into a world of imagination, exploration, embodiment and emotion. In Luci's world, water indeed becomes a material agent, or a form of 'vibrant matter' (Bennett 2010). The chapter is deliberately reflective and celebrates narrative's 'intrinsic incompleteness' and the role of the audience in bringing iterative meanings and versions to the texts in an ongoing way.

Antony Lyons (Chapter 3) brings to life the interweaving of the sensory, the experienced, as well as the imagined and thought about, and implicitly illustrates the creative practitioner as a kind of interlocutor, taking on a role of making novel connections and facilitating new interactions with landscape and sense of place. The creative process involves bringing together materials, imagery, poetry and ideas into spaces, both drawing attention to the existing qualities of those spaces as well as questioning and unsettling them. In one example, a Portuguese wash house becomes the space of bringing together. Here, natural materials from the surrounding landscape, imbued with cultural significances, are placed in the wash basins of this mundane yet important communal space. A performative reading of an old Irish poem in which the orator becomes personified as 'the salmon in the stream' or the 'waves upon the sea' is translated into Portuguese. Through this combination, we see a reassembling of socio-natures in a way that is both deeply embedded as well as novel. There is a fluidity of identification, and an incorporation of the lived and embodied experience of being in space, reflection on identity and community through new – and newly unearthed – connections. Lyons draws inspiration from sources that include cinematic and poetic works, and from the writings of Ivan Illich. We can see here how the incorporation of embodied, personal, cutural and fluid elements could relate to a notion of 'dwelling', understood in Ivan Illich's terms as counter to the alienation and abstraction of the commercial and modern view of water.

In Chapter 4, Bakewell, Liguori and Wilson focus on the use of creative methods within an academic project focused on collaborative water management. Taking inspiration from an historical community conflict resolution method (*La Rasgioni*) from the Gallura region of Sardinia, and combining this with understandings of theatrical performance, music, storytelling and visual art, they recreated a community courtroom in the Fens region of the UK designed to encourage both community conversation and community cohesion with regards to water management. They discover that collective meals play an enormous role in both conviviality and conversation, and that the performative, humorous and enjoyable character of the events also had the effect of encouraging people to linger and converse. Though these methods could be applied to many different issues, the suggestion that creativity be brought into the spheres of water management and policy expands the notion of water governance through creating new possibilities for creative conversation. The question remains whether this process can occur in the same way outside of an academic project and whether and how this might be of long-term benefit to a community and its water management.

Part II Becoming water bodies

This section brings together chapters that explore embodied and multisensory modes of experiencing water. They all focus on how bodies engage with one particular form of water: the sea. Through embodied accounts they produce ways of knowing water that incorporate tacit, memory-based and emotional aspects, deeply tied to sense of self and identities: as swimmer, as artist, as surfer. These identities elicit an ecological perspective to varying degrees and consider the affective component that watery encounters hold for particular individuals. Our title for this section gestures to work on performance/performativity that argues for ideas about pre-cognitive affects and emergent relations as the ontological basis of our experience. The authors express the experience of 'becoming' via a particular type of attunement to the water.

We start this section with Ronan Foley's (Chapter 5) ethnographic and auto-ethnographic discussions and reflections of the intimate, embodied and relational aspects of sea swimming across the life course. Embodied engagements with the sea, and a culture of sea swimming that spans age, gender, culture, community (as well as risk), provide a way for individuals to discover and rediscover a sense of themselves in relation to the world over time. Swims are associated with moments in time, with life events, with warm connection and relationships with other people, a sense of belonging in place and a sense of relationship with water that is both beautiful and unpredictable, frightening and a challenge, a test in which the swimmer finds themselves rejuvenated in body and spirit. There is a sense of ritual about the swims, an undressing and dressing, a dip in the cold winter sea or a longer more meditative swim in the warm clear waters off a Greek Island. Swimming itself becomes an everyday act of creativity (Hallam and Ingold 2007), as swimmers and water co-constitute each other and come into relationship. There is a sense of change over the life-course and the shifting and altering of the relational web in which particular swimming spots, people, places and the individual intersect at different points in time.

While Foley reflects on the relational sense of belonging and being brought about by swimming in the sea, in Chapter 6 Anderson and Stoodley explore a similar but different interaction with the sea and consider how surfing might be seen as a 'creative compulsion'. Drawing on ethnographic and auto-ethnographic research with surfers and surfing they consider how the 'liquid gold' found in the moment when geomorphological conditions, wind conditions and crucially swell perfectly combine to produce surface waves on the sea, along with the body of a surfer, result in the momentary creation of an identity as surfer. The authors discuss how this sense of identity is fleeting, tied into the precise moment the elements come together, and that this feeling compels the individual to seek the same combination again and again; it is only in the moment of becoming surfer that they 'feel right'. Anderson and Stoodley compare the moment of becoming surfer to the compulsion of an artist, carving a wave rather than producing a sculpture, painting or other traditional artefact. Coastal surfing is an activity reliant on geographical, hydrological and climatic conditions, often making surfers inextricably aware of and connected to the water (and land and air) environment. The chapter outlines how this awareness motivates activism to protect local surf spots or take wider action to reduce environmental damage to beach and ocean, as well as to designate 'World Surf Reserves' to be protected more formally.

An essential element in the act of surfing is the nature of waves. In Chapter 7, John Hartley explores just this, through sea swimming, drawing and hacked recording technologies. Intricate line drawings form sets of waves, meeting and influencing each other in Hartley's creative response to re-thinking the 'emblemata' of knowledge from rock-like hard facts to fluid waves. We have a sense of the struggle with the definitiveness of so much modern techno-scientific 'knowledge' and the realities of life that are continually in flux and change. Hartley's explorations feel like an intuitive attempt at casting off an imposed fixity, while still embracing a desire to capture some of the flow in representation. We are taken on a journey towards how it might be to think like a wave, or as a wave, to be more aware of our own wave-producing capacities through our interaction with water than we are of the hard and distinct boundaries of human bodies. This feels an inclination towards Neimanis' call to recognise ourselves as part of a flow rather than discrete beings.

Hartley goes on to play with other assemblages in a quest for expressing the feelings associated with swimming in the sea via hacking of simple, cheap technology – old digital cameras in mason jars – for recording swims. The selection of outmoded technology is also seen as part of a wave and through exploring this notion, we are enabled to see the meaning production associated with objects and an objectification of water, in a new light. If an object remains the same but changes in meaning according to cultural, socio-technical waves, then a water body also can simultaneously remain the same and change through the interaction not only of other bodies but of ideas.

The third section of the book aims to bring together chapters that illustrate ways in which knowledge about water can be produced or represented. This involves questioning representations of water such as diagrams of the hydrological cycle, that might be quite familiar to us, as well as delving into a non-visual world of representation and considering how sound might provide a different entry point to getting to know water. These chapters state the case for water knowledge as being 'unfinished' and emergent in relations between humans and water. Further to this, they explore how relationships are mediated through particularly disciplinary lenses, through technologies and through discourses of morality. They show how knowledge can be co-produced with an 'other' outside of the usual knowledge-making confines of academic convention, either in participatory workshops, with more-than-human species and between creators and audiences.

To begin the section, Farnum, Thompson and MacDougal (Chapter 8) take one of the most common and recognisable representations of water, the diagrammatic representation of the hydrological cycle that was developed in the 1930s, which has been a staple of geography school texts ever since. The authors begin with a project that brought together an artist, a social scientist and a hydrologist and then via workshops and discussions looked for ways to re-represent the hydrological cycle as a hydrosocial cycle. The results of their workshops were diverse terms associated with and ways of looking at water, and creative and interesting ways of representing these in drawings and diagrams. The artist on the project sought to facilitate discussions through designing a hydrosocial spiral with possibilities for modification through individual panels. They highlight how this process was productive accross different settings for learning and reflecting on human–water relationships. Interdisciplinary engagement with stakeholders enabled a wider conception and adaptation of the tool than was initially imagined, whereby the process became more important than the final artwork/tool.

In Chapter 9 Rob St John describes his process, as an artist and cultural geographer, of using sound recording technologies to capture the underwater soundscapes and materialities of water, as well as the creative, affective force of water as an agent. He offers this type of approach as one way to move beyond conventional understandings of sound (preciseness, isolation, clarity, referentiality) and promote a more attentive listening, with associated intensities and searches for meaning, which bring us closer to research exploring more-than-human worlds and socio-ecological concerns. St John adopts an auto-ethnographic approach, building on this through his own practice as an artist through vignettes. He outlines a range of technologically mediated encounters with soundscapes novelly bringing these together with more-than-representational debates to illustrate how creative methods might be used for 'rendering things strange' by de-privileging human experience and promoting other river ecologies and a 'being-with' philosophy of valuing life at all scales.

While St John's chapter involves listening to the sounds made by water and seeking to understand relationships through this, Meisch (Chapter 10) instead turns his listening ear to religious poets of the past. He argues that looking at historical literatures and biblical creative forms such as hymns that reference climate change events such as the Little Ice Age can help us to think differently about anthropocenic and/or anthropogenic climate change, by providing social and moral contexts to extreme weather that is 'other' to our own, creating empathic and creative responses in the reader that may cause us to reflect upon our own socio-technical regimes for thinking about human–water relationships. Meisch reflects on the ways in which religion as well as literature have in the past both reflected sentiments about weather and climatic conditions, such as through beseeching god for rain in times of drought, but also how these moralistic sentiments can be seen paralelled in modern, albeit scientifically driven discourses around climate change. Suggesting this parallel, and providing this critical distance through looking at historical literature, has the possibility of shedding some light on contemporary moral and ethical debates as they relate to climate change, opening up space for perhaps a wider look at this moment in contemporary thinking and the implications for wider society.

Part IV When water disrupts: water as agent and co-constitutor of place and culture

Our final section gives us the most detailed discussion of water politics. The chapters show how water can act on a community as a creative force for change and how it can encourage tactical adaptations within everyday contexts due to too much or not enough access to it. The chapters highlight how when water acts as a disruptive element to the status quo new constellations of people–place–politics emerge. These chapters show how water can be agential and how people can resist through their water practices the usual behaviours or power relations associated with 'modern' or 'Big' water. Water can disrupt in social, cultural and political spheres, in the embodied micro-politics of social practices, for example in the home, or through state-level politics via conflict in public arenas. Challenges in the form of small-scale, personal reflections or community-level conflict over the status quo enable a re-thinking of contemporary water systems with degrees of opportunity for change.

In Chapter 11, Birkinshaw examines the social and political variabilities observable in informal water supply for domestic use in the Sangham Vihar region of Delhi. Here we find an intricate system involving formal and informal mechanisms for water access that are tightly linked to local politics.

The materiality of water in this context is also relational, disruptive and unruly within the socio-political system, which is made more vulnerable through environmental/hydrological uncertainty. Groundwater is depleted, socio-political arrangements and intricacies see water supply operating in mysterious ways and a population reliant on various combinations of tubewell and tanker improvise through obtaining enough potable water to survive. Water is not simply water within this mesh. Groundwater extracted at depth is very hard, causing damage to the technologies used for extraction, as well as having implications for human health. We are reminded in every interaction of the material as well as the socio-political aspects of water flows in an urban space.

While the example from Delhi illustrates the continuous lived reality of unreliable access to water for domestic needs, Hoolohan and Browne (Chapter 12) discuss what happens when the relationship with an easy supply of water in a UK context is altered through camping music festivals and a rivercare project. Alternative types of encounters with water and water supply can create a pause in the often, unthought relationship between water use and its origins that has evolved through what they term 'Big Water' systems of provision. Though it is difficult for individuals to break from the lock-in, it forces us to consider different policy options that move away from placing responsibility on the individual water consumer and towards a more holistic approach. Specifically, it is the way that 'talk' happens and we socially change the norms through experimentation in practice and interactions with others sharing the experience. They propose such interventions might 'cultivate more convivial ecological sensitivities, capabilities and practices' (Hoolihan 2016) and embrace creative experimentation, advocating work that seeks transformative encounters with water where the site could be the home, the body, the community as can be found in this volume. The chapter provides a critique of 'Big Water' as the only option within neoliberal contexts, instead suggesting this is the problem.

The perception of what we know about water and indeed climate change come to the fore again in Chapter 13. Here, Bonnelykke takes us on a journey to Kiribati where we try to 'think like water moves'. Kiribati has attained a kind of fame due to the discourse around climate change, as these islands look due to be submerged by the sea with even modest sea-level rise. Oral histories and films have begun to emerge depicting the people of these islands and indicating the injustice of losing their homes due to a human-induced process of which they have had a very small part. Through the chapter a dialectic emerges whereby inhabitants accept both continuity and change in understandings of life on islands that themselves are part of processes of both change and continuity; islands that emerged from the sea may again return to the sea. The attitude of inhabitants is sometimes very different to how outsiders bring news of anthropogenic climate change. The human interventions on the islands over time have made use of existing conditions and altered them through different patterns of growing and fishing, for instance, so that new ways of doing and being also become part of the understanding of relationships with land and sea. This is a clear reminder of the situatedness of cultural understandings and feelings as well as the globalisation of discourses and the implications and interactions that this brings. The Kiribati Islanders' creation stories and relationship with the sea is one of accepting change, the incursion of water into life and a balance of living with duality. From the water emerges life and livelihoods, and yet the water may also take homes and belongings, and even all of the islands. Living with this duality of water involves in Bonnelykke's words 'thinking like water moves'.

To conclude the collection, we take a look at two watersheds in Oregon, USA. Rudestam, reminds us again of our beingness, of affect and emotion and of how 'world-making' is a relational process. Here we are given insight into water-management practices in two watershed basins in Oregon, USA. The two basins, while close together, have different physical geographies (one river is 'flashy', prone to floods and vulnerable to droughts, while the other is a stable watercourse), as well as different human geographies (a single organised tribe versus multiple tribes). Both basins have, as Rudestam points out, been subject to colonial and capitalist expansion, indicating a strong political bias from a colonial government towards irrigation above all other water uses. Adding to this mixture, the effects of new knowledge of, as well as the material effects of climate change and the effects of limiting water in the river through damming and irrigation on fish species (particularly ones that are endangered), creates in one basin a situation of conflict, and in the other (by many accounts as a direct effect of the conflict in the first basin) a situation of collaborative water governance. Matters of water governance are affected deeply by emotion, whether it is the emotion of fear that a governance situation may turn out like another example, or whether it is the emotion that experiences of and with water engender that lead to action or inaction in different contexts. This final chapter returns to the driving force behind bringing this collection together: the argument for making space for creative ways of thinking and being in relation to water.

Through bringing together this set of distinct and complementary chapters, we have sought to foreground creative methods and practice in multi-disciplinary water projects as well as independent creative practice that offers different ways of knowing, relating to and understanding water. These have potential to intervene and disrupt current 'resource-' focused frameworks prevalent in academia, industry and policy. The creative projects outlined in our chapters represent different and alternate ways of knowing water often at odds with the 'resource' discourse, which has led to our current ecologically fragile position. While economic priorities in water use prevail, there are rising anxieties about environmental damage and attempts to valorise eco- and aquatic systems, however, these are often translated through the same techno-managerial frameworks into weak or inadequately reinforced sustainability policies and legal and material regulations (Strang 2016). Culturally diverse ways of knowing and engaging with water have been overlaid with a more homogenous view of water that 'assumes water policy is primarily the remit of science and scientific practice' (Linton 2010; Strang 2016, 10). In response to this, we echo the plea that in order to develop more genuinely equitable and sustainable water practices, 'we should start from an analytic perspective substantially different from where we are now' (Strang 2016, 1). The collection is a concerted effort to re-value and re-frame our relationships with water through creative experiments.

There is a temptation to try to justify the value of this type of work or, in academic terms, its 'real-world' impact, through outlining in practical terms how it can inform policy, planning and industry – i.e. the water resource sector. We suggest, instead, that the (political) intervention of this work is at the level of micro-politics. Drawing on the work of Strang and others (2014, 2016; Krause and Strang 2016), we offer a framework for a 'water ethics' inspired by creative approaches broadly, by the chapters enclosed, and outlined in full within our concluding chapter at the end of the book. We wish to resist the pull of the 'modern water' (Linton 2010) rationale with its scientific, technical and economic framings as it ripples/echoes out through other neoliberal contexts like academia, to put water, work on water, *or* creativity, and work on creativity, again in the service of human power relationships, institutions and practices that dictate how we should value water. We will not try to assimilate the contribution of these chapters into an instrumental language that turns them into 'services' in order to make them relevant to resource managers and planners. Instead, our chapters simply offer a pluralism for thinking about human–water relationships that incorporates experiential, embodied, affective, emotional, informal and ecological encounters with different forms of water, related meanings and values.

References

Abukhater, A. (2013) *Water as a Catalyst for Peace*. Oxford and New York: Routledge.

Alberti, B. (2014) How Does Water Mean? *Archaeological Dialogues*, 21(2), 160–162.

Bakker, K. (2012) Water Security: Research Challenges and Opportunities. *Science*, 337(6097), 914–915.

Barrett, E. and Bolt, B. (eds) (2014) *Practice as Research: Approaches to Creative Arts Enquiry*. London: I.B. Tauris.

Bastian, M., Jones, O. and Moore, N. (eds) (2017) *Participatory Research in More-Than-Human Worlds* (Vol. 67). Oxford: Routledge.

Bennett, J. (2010) *Vibrant Matter: A Political Ecology of Things*. London: Duke University Press.

Cáceres, Berta. (2015) *Goldman Environmental Prize, Honduras* [Film]. San Francisco, CA: Goldman Environmental Prize. www.goldmanprize.org/recipient/berta-caceres.

Callon, M. (1986) Some Elements of a Sociology of Translation: Domestication of the Scallops and the Fishermen. In Law, J. (ed.) *Power, Action and Belief: A New Sociology of Knowledge*. London: Routledge.

Carson, R. (1951) *The Sea Around Us*. New York: Oxford University Press.

Carson, R. and Hubbell, S. (1998 [1955]) *The Edge of the Sea*. New York: Houghton Mifflin Harcourt.

Castree, N. and Braun, B. (2001) *Social Nature Theory, Practice, and Politics*. Malden, MA: Blackwell.

Crutzen P.J. (2006) The 'Anthropocene'. In Ehlers, E. and Krafft, T. (eds), *Earth System Science in the Anthropocene*. Berlin, Heidelberg: Springer, pp. 13–19.

Davies, M. (2014) Don't Water Down Your Theory: Why We Should All Embrace Materiality But Not Material Determinism. *Archaeological Dialogues*, 21(2), 153–177.

Deakin, R. 2000. *Waterlog: A Swimmer's Journey Through Britain*. London: Random House.

Gandy, M. 2014. *The Fabric of Space: Water, Modernity, and the Urban Imagination*. Michigan: MIT Press.

Glaveanu, V.P. (2015) Creativity as a Sociocultural Act. *Journal of Creative Behavior*, 49(3), 165–180.

Gleick, P.H. (2014) Water, Drought, Climate Change, and Conflict in Syria. *Weather, Climate, and Society*, 6(3), 331–340.

Gooley T. (2016) *How To Read Water: Clues, Signs & Patterns from Puddles to the Sea*. London: Sceptre.

Guyotte, K.W., Sochacka, N.W., Costantino, T.E., Walther, J. and Kellam, N.N. (2014) STEAM as Social Practice: Cultivating Creativity in Transdisciplinary Spaces. *Art Education*, 67(6), 12–19.

Hallam, E. and Ingold, T. (2007) *Creativity and Cultural Improvisation*. London: Bloomsbury.

Hanasz, P. (2017) Transboundary Water Governance and International Actors in South Asia: The Ganges-Brahmaputra-Meghna Basin. London: Routledge.

Haraway, D.J. (2016) *Staying with the Trouble: Making Kin in the Chthulucene*. Durham, NC: Duke University Press.

Hawkins, H. (2015) Creative Geographic Methods: Knowing, Representing, Intervening. *On Composing Place and Page: Cultural Geographies*, 22(2), 247–268.

Hawkins, H., Marston, S.A., Ingram, M. and Straughan, E. (2015) The Art of Socio-ecological Transformation. *Annals of the Association of American Geographers*, 105(2), 331–341.

Hoolohan, C. (2016) *Reframing Water Efficiency: Towards Interventions that Reconfigure the Shared and Collective Aspects of Everyday Water Use*. Thesis submitted to the University of Manchester for the degree of Doctor of Philosophy (PhD) in the Faculty of Science and Engineering.

Illich, I. (1985) *H2O and the Waters of Forgetfulness: Reflections on the Historicity of 'Stuff'*. Pennsylvania: Heyday Books.

Jha, A. (2015) *The Water Book*. London: Headline.

Joas, H. (1996) *The Creativity of Action*. Cambridge: Polity Press.

Kaika, M. (2005) *City of Flows: Modernity, Nature, and the City*. London: Routledge.

Kindon, S., Pain, R. and Kesby, M. (eds) 2007. *Participatory Action Research Approaches and Methods: Connecting People, Participation and Place*. Oxford: Routledge.

Krause, F. and Strang, V. (2016) Thinking Relationships through Water. *Society and Natural Resources: An International Journal*, 29(6), 633–638.

Law, J. (2009) Actor Network Theory and Material Semiotics. In Turner, B. (ed.), *The New Blackwell Companion to Social Theory*. Chichester, UK: Wiley-Blackwell, pp. 141–158.

Linton, J. (2010) *What Is Water? The History of a Modern Abstraction*. Vancouver, BC: UBC Press.

MacGregor, D. (2017) *Anishnabek Gilnendaasowin, Research and Water Governance*. Royal Geographical Society Annual International Conference, RGS, London.

Mehta, L. (2010) The Social Construction of Scarcity: The Case of Water in Western India. In Peet, R., Robbins, P. and Watts, M. (eds), *Global Political Ecology*. London: Routledge, pp. 371–386.

Mirumachi, N. (2015) *Transboundary Water Politics in the Developing World*. London: Routledge.

Morton, T. (2017) *Humankind: Solidarity with Non-Human People*. London: Verso Books.

Naess, A. (1973) The Shallow and the Deep, Long-Range Ecology Movement. A Summary. *Inquiry*, 16(1–4), 95–100.

Neimanis, A. (2017) *Bodies of Water: Posthuman Feminist Phenomenology*. London: Bloomsbury.

Pearce, F. (2006) *When the Rivers Run Dry: Water – The Defining Crisis of the Twenty-First Century*. Boston: Beacon Press.

Reisner, M. (2017) *Cadillac Desert: The American West and its Disappearing Water*. New York: Penguin.

Strang, V. (2004) *The Meaning of Water*. Oxford: Berg.

Strang, V. (2010) Water, Culture and Power: Anthropological Perspectives from 'Down Under'. *Insights, Journal of the Institute of Advanced Study*, 3(14), 2–26.

Strang, V. (2014) Fluid Consistencies: Material Relationality in Human Engagements with Water. *Archaeological Dialogues*, 21(2), 133–150.

Strang, V. (2016) Re-Imagined Communities: A New Ethical Approach to Water Policy. In Conca, K. and Weinthal, E. (eds), *The Oxford Handbook of Water Politics and Policy*. Oxford, New York: Oxford University Press, pp. 142–166.

Thistlewood, D. (ed.) (1995) *Joseph Beuys: Diverging Critiques*. Liverpool: Liverpool University Press and Tate Gallery Liverpool.

Tisdall, D. (1974) *Art Into Society, Society Into Art*. London: ICA.

Welsh, M. (2014) Resilience and Responsibility: Governing Uncertainty in a Complex World. *Geographical Journal*, 180(1), 15–26.

World Economic Forum. (2017) *The Global Risks Report*, 12th edition. www3. weforum.org/docs/GRR17_Report_web.pdf.

Xie, L. and Shaofeng, J. (2017) *China's International Transboundary Rivers*. London: Routledge.

Part I

Fluid processes

Creative research with water

Chapter 1

Water power

Creativity and the unlocking of community knowledge

Loraine Leeson

The flow of the River Thames has been the lifeblood of London over many centuries. Since the first Bronze Age settlers it has provided sustenance, transport, trade, work and pleasure for a population that now exceeds 8 million. Communities grew up along the river to the east of the city to service its developing trade, but while this area was a hub of wealth generation, it has also seen the city's greatest poverty. A focus point for new immigrants, it became home to a range of different cultural groupings, from the Huguenots in the seventeenth century onwards. Many entered with the trade ships, gravitating toward the work opportunities and cheaper living of this industrial quarter or aiming to join others from their own cultural backgrounds. Despite the poverty, however, it would be a mistake to see East Londoners as victims. Necessity drove these riverside communities to become highly organised, and it is their determination and resilience that has led to this current urban territory of astonishing energy, diversity and culture. It is not surprising therefore that its local inhabitants still recognise a role for this river in sustaining their lives. Indeed, when in 2007 members of the Geezers Club in an AgeUK East London centre were asked about technological needs for a research project, they looked to the river for solutions.[1]

Geezer power

This question was nevertheless put to this group, not by a technologist, but by myself as an artist commissioned by researchers who recognised the transformational potential of arts development methods. The research project in question, *Democratising Technology*,[2] was led by an interdisciplinary team at Queen Mary University of London, who were exploring whether a generative, open-ended form of engagement between communities and technology could be produced and brought to bear on the design of society and its tools. The research team had been examining how the experience of older people was not only being excluded from the development of new technologies, but often left this age group victim to the technological design and control of others. They commissioned three artists, of which I was one, to work with older people's groups around these ideas.

There is increasing recognition of the effectiveness of the arts in working alongside other disciplines to address issues of social relevance. While the arts can rarely implement change on their own, they are adept at communicating and consolidating ideas in a way that facilitates engagement, and particularly so when this includes drawing out the creativity of others. Creativity can be seen as the process through which an idea is 'made real', where energy interacts with matter to make something entirely new, whether that is a planet, a culinary dish or a painting. In using this method it helps to have some experience of the creative process as well as familiarity with the art of 'not knowing' – to stay with ideas as they emerge and change, then discover the meanings as they manifest. This is not a pre-requisite of the arts, but it is a process in which artists are well schooled and learn to manage productively.

From my experience of working as an artist with communities for almost four decades, I have discovered that if there is a need or issue to be addressed, then those most affected will hold important knowledge and may also already be working in some way to resolve it. Processes of creative facilitation can help people articulate what they know and communicate it in forms that are accessible to a wider audience. This approach nevertheless also requires some relinquishing of the notion of individual creation that has for so long been collapsed in public perception with the notion of the artist,[3] and embracing the shared creativity of collaboration. Not all artists choose or are able to work in this way since it requires a range of skills and interests beyond the artistic. However it is an approach that has slowly developed as a professional pathway since the 'community arts' movement of the 1970s and 1980a. In the last 20 years a new movement of 'socially engaged' art has emerged that is not only accepted, but now even promoted by major art institutions. This is a chequered history that I explore further in my recent book *Art: Process: Change* (2017) that also delves into key methodologies employed in this endeavour.

The arts commissions offered by the research project were managed by SPACE[4] with results to be exhibited in their gallery six weeks later. The assumption was that the 'research' had already been carried out by the academics and it was the artists' role to 'respond' to this. While it was encouraging to see the social role of the arts recognised through the commissions, it was familiar but frustrating to encounter the assumption that the artistic process was essentially only about production. Through my art practice to date I had come to understand how research and production are interdependent. The creative process enacted through practice develops new knowledge, which is by its nature innovative and delivers substance that can then be examined. A more fruitful approach would have been to commence with the art, which would then have provided material for the academic researchers.

The six-week turnaround also had to encompass the building of the relationships necessary for meaningful community interaction, which together with the necessity of re-commencing research with this group, made for a tight call. The importance of long-term engagement with participants in socially engaged projects is becoming increasingly understood and I do not normally take on short-term commissions for this reason. However interest in the topic backed by the experience and support of the gallery and goodwill of the research team, suggested that the attempt would be worthwhile. It proved to be a risk worth taking, since it opened up a new creative pathway that generated a collective energy that continued to drive it under its own steam. Ten years on, the project is still going strong supported by funding that I have patched together as we progressed. During this time we have enlisted a professional engineer and a range of partners, created exhibitions, investigated how turbines might function on the Thames flood barrier, tested a small-scale tidal turbine in its central London reaches, run schools' workshops, engaged in interdisciplinary presentations and international virtual communications and most recently created a floating water wheel to aerate the water and support fish and wildlife in a Thames tidal basin (Figure 1.1). I will, however, return to the beginning of this process to examine more closely the relationship, understanding and engagement of this particular community with the power of its river.

I met the Geezers at one of their regular meetings at an AgeUK centre in Bow. The self-named Geezers Club had been established in 2006 to mitigate the effects of isolation and loneliness on older men and offered members access to social activities, outings and talks by outside professionals. These were often on health-related topics, however, and it was clear that the group were delighted by the opportunity to work on something that drew on their skills and experience. During the project the group placed immense value on the fact that the activity was not just for its own sake or to pass the time, and one through which they could pass on their own accumulated knowledge to future generations. This dovetailed

Figure 1.1 Visualisation of tidal turbines on the Thames Barrier. *The Not Quite Yet*, SPACE, London, 25 January–29 February 2008 (© Loraine Leeson).

well with my position as an artist interested in enabling the impact of otherwise marginalised voices to enter public discourse and feed into social change.

My opening question to this group of working-class men was on what technological developments each felt might best assist them or their communities in the future. I thought this might have been answered with ideas for gadgets or domestic aids, however the group had bigger ideas, formulated out of their lived experience in one of the country's poorest boroughs that also borders one of its largest tidal rivers. 'When electricity prices prevent older people from heating their homes, and the Thames is just down the road, why aren't we using it to power our community?' asked one member of the group. It turned out that many individuals in this area could not afford to live in the sheltered accommodation they needed due to service charges inflated by the high cost of energy. Yet in this group was an ex-steam engineer and others with practical skills who could see the potential that the 'powers that be' were missing. Many remembered how, decades before, tidal and wave power had been in the news but recently they had only heard of wind technology. Others recalled how in previous centuries a water wheel had been attached to London Bridge, while a nearby heritage site housed the remains of the world's oldest and largest tidal mill. By the end of the first session the whole group were keen to focus on nothing less than harnessing the tidal power of the River Thames. While I had no experience or particular knowledge of these issues, I pledged that I would help take the Geezers as far along this route as it was possible for us to go.

We began by visiting the mill to learn how its now rusted wheels had once been used to turn stones for grinding grain, and also that current volunteers had plans for a new turbine that would bring it back into use. I conduct most of my practice through an arts organisation cSPACE,[5] which was based at that time in the University of East London and here I discovered the Director of Sustainability, who advised us further. It seemed that funding for tidal technology had been severely reduced in the 1980s, with later development of renewable power sources focused mainly on wind energy. There were no readily available designs for turbines that could respond to the river's ebb and flow and so, under his guidance, the group organised community transport to look at locally sited wind turbines that could most easily be adapted for underwater use. A visit to the Thames Barrier also revealed a suitable ready-made barrage for potential turbine installation. From visual materials gathered in our research I was able to create a large-scale photomontage of how turbines might function in this location. The group's new knowledge coupled with their understanding of its potential benefits for the lives of local people also made them highly effective advocates of the sustainability argument. For the exhibition therefore, I conducted video interviews with its members to accompany the photo-visualisation. These were projected at a large scale to lend a weight of authority to the views of the speakers. The impact of this installation on gallery visitors was reflected in significant local press coverage. Despite little experience of public speaking, eight members of the group presented the project to great acclaim at *On the Margins of Technology*, the symposium that accompanied the exhibition. This attention was ironically much to do with the very nature of group members' senior status, which caught people's imaginations, turning on its head their initially marginalised position.

Active energy

The creative energy generated by the project, from that point entitled *Active Energy*, gained its own momentum, and at the end of the commission we all felt that it was not possible to halt the work there. After the exhibition I found a small amount of funding to equip the Geezers with a laptop and other equipment that would allow its members to learn the skills to conduct online research and share their findings. Group members were enormously engaged in the potential of their idea, which tapped into their existing skills and interests. Unprompted, they began to draft new turbine designs and debated how these would work. Engineering expertise presented itself in the form of Toby Borland, a highly creative mechanical engineer who ran a prototyping laboratory at University of East London, and Professor Stephen Dodds, renowned for his development of the control system for the European Space Commission. Both gave freely of their time and knowledge out of interest in the project. SPACE arts organisation, which had managed the original arts commission, re-joined the project for similar reasons, raising funds to support intergenerational work with a local school as well as continuation of the Geezers' work on tidal energy. Through this collaboration I facilitated Toby Borland to lead the school workshops, assisted by Stephen Dodds, while previously isolated older men from the Geezers Club now found themselves mentoring underachieving boys. At the school's request the work focused on wind power, and so the young people learned about aerodynamics and tested their designs in a makeshift wind tunnel. The best design was then used for a wind-driven lightwork for the roof of the AgeUK centre, which rotated to spell out 'geezerpower' (Figure 1.2).

In the meantime Professor Ann Light, who had been involved in the initial research for *Democratising Technology*, returned with an opportunity for the Geezers to contribute to *Participants United*, a research event at University of Central Lancashire, which was exploring constituent processes involved in effective community engagement. In this way the Geezers were able to extend their new-found knowledge by sharing it with other community initiatives. This exchange between community peers from different locations has now been realised by this group in a number of different contexts. In 2012 an opportunity arose for international dialogue with a seniors' group in Pittsburgh through an

Figure 1.2 Wind turbine on Age UK centre, Bow, February 2010 (photo © Loraine Leeson).

artist residency I was conducting there at that time. Connecting via Skype, a first for all concerned, the Geezers inspired a group of women at Northside Seniors to conduct their own project on a topic of their choice. The results were exhibited through a six-projector installation at the Mattress Factory museum that high-lighted the significance of older people's potential contribution to contemporary social issues. It is no coincidence that, given the opportunity, these two groups of working-class elders set out to tackle some of the key issues of their respective nations. The strength of the desire of each to leave a legacy for the next generation and the experience that lay behind it ably contradicted those who write off the value of older people's potential contribution to society or think their offering would not be relevant to the present day. The Geezers' recognition of London's river as both an historical, and currently much-needed source of sustainable energy for the capital, has been well in advance of the plans of government agencies or energy companies and brought about by thinking specifically from the perspective of local needs outwards.

Meanwhile design work on a tidal turbine continued in London with ongoing support from engineer Toby Borland. The Geezers developed designs at the University of East London's prototyping laboratory, trying them out in a

Figure 1.3 Testing turbine efficiency at the University of East London, March 2010 (photo © Loraine Leeson).

specialist water tank (Figure 1.3). A suitable riverside site that could support the final prototype chosen had to be found in the Thames, and the owner of a barge that functioned as a bar close to the Houses of Parliament offered use of his vessel. Although the testing demonstrated more work to be done, the process of development identified the device as the first small-scale turbine suitable for use on tidal rivers. Its production from low-cost and recycled materials made the design eminently adaptable for use in situations where cost would be an issue such as in developing countries. All the designs were created to be open source and posted on the *Active Energy* website[6] for others to access.

Active Energy typifies the organic way in which such projects can develop and gain longevity when they are rooted in community and not subject to overarching commissioning constraints. Despite frequently lacking the benefits of advance funding, work such as this is able to respond to need and opportunity. While its central aim has been to support the older communities of East London through inexpensively produced sustainable energy, *Active Energy* has also been able to extend its reach into schools, across the Atlantic to empower other seniors' groups, and more recently to the north of England to engage with the work of Canal Connections, a community initiative based in Leeds. Development of the turbine also led to an additional, parallel, two-year project with the owner of the barge where the Geezers tested their turbine. This explored the dearth of wildlife habitat along the river's urban reaches, where historic marshlands had been transformed into shored-up concrete banks to enable sufficient depth for river traffic. *Lambeth Floating Marsh*[7] experimented with the construction of reed beds along the hull of a Thames barge to provide an experimental environment where microorganisms and invertebrates could breed and support the river's food chain. Images of these organisms were then projected along the embankment to bring the issues to public attention.

While the river first identified by participants in the project was the Thames, the group instead shifted its focus to one of its tidal basins through inclusion in *Hydrocitizenship*.[8] This research project, involving 15 researchers from 9 universities, had for three years investigated and promoted the relationship of communities and water. In common with the research that initiated *Active Energy*, *Hydrocitizenship* started from an understanding of the value of the arts in community-based research, although in this instance has involved artists from the outset. In the case of *Active Energy* we did not need to set out to foster a relationship between citizens and water, since that already existed in the project, but were able to extend our remit to address further issues affecting the tidal reaches of the Lea Valley as it joins the Thames. Through partnership with Thames21,[9] which aims to rebuild the relationship between communities and their rivers while restoring river health, the group came to understand some of the concerns with pollution. This was particularly poignant for those Geezers who in their youth used to fish in the River Lea.

Under certain weather conditions sewers overflow into the river and the bacteria from the effluent feeds microorganisms, which then take up the oxygen in the water so that fish suffocate. Through workshops with Toby, the engineer who had stayed with the project since its early days, we together worked out a plan to use the river's flow to drive an aerator that would pump oxygen into the water. An excellent site for this proved to be close to the nearby historic Three Mills site, knowledge of which had in part informed the Geezers' understanding of the power of water at the commencement of the project. Run-off from the millpond could be used to drive a wheel as it emptied with the falling tide. Since the tidal range is quite extensive here, and at its lowest ebb less than a metre in depth, we arrived at the idea of a floating water wheel rather than a turbine, that could rise and fall with the tide (Figure 1.4). Due to the permissions required for placing objects in rivers, it was only possible to try out the wheel for a six week period, during which time it functioned well. Plans are now under-way to install it at a new site in the Queen Elizabeth Olympic Park, where it can become a focus for workshops and events to highlight the use of rivers as sources of renewable energy. In the meantime this evolving project, patched together with little funding and much goodwill, was honoured with RegenSW's first *Arts and Green Energy Award*.[10] This regeneration agency had hitherto focused on industry achievement, and the addition of this award for art projects has been further recognition of the significance of creative practice in this field.

Figure 1.4 Water wheel at Three Mills, May 2017 (photo © Loraine Leeson).

Art, the river and social change

I did not set out to do work about the Thames. However, since my early excursions into community-based art practice, the concerns, interest and hopes of the people I have worked with, particularly in East London, have demonstrated the strength of local people's connection with their tidal river. As such it has often become central to my work and a location to which I have repeatedly returned. This has encompassed the campaigning in the 1980s to ensure that the London Docklands continued to support its communities in the *Docklands Community Poster Project*, and the 'rainbow river' of the stand against racism on the Isle of Dogs in *Celebrating the Difference* during the 1990s. In the new millennium the identity of longstanding and migrant riverside communities was explored in *Precious Places*, following which young people celebrated their waterside neighbourhood in the *Young Person's Guide to the Royal Docks*. More recently *Lambeth Floating Marsh*, described above, addressed the issue of microorganism habitat in the river's urban reaches.[11] The content of all these projects has been led by their participants and collaborating partners, and as a result I have constantly found myself drawn back to the river for reasons of its historic, symbolic or transformative power. Ultimately the significance of water did not have to be introduced to the Geezers, rather they were asked a question, the answer to which drew on their local knowledge as a riverside community.

Although I provided artistic leadership and management for these projects, and certainly choose areas of broad concern that interest me, it has been the participants and collaborative partners who have offered the direction and purpose. In *Active Energy* it can be seen how this has then been developed with people from other disciplines and walks of life to bring substance to the ideas. This project's informal team of individuals and groups with diverse interests have now become used to delivering group presentations via an interdisciplinary panel involving various combinations of artist, engineer, social scientist and participating Geezers. At one point, intrigued by the vortex of creative energy generated through this mix, we all commented on what it meant to our respective interest or discipline. The Geezers' goal was clearly to improve the lives of the older population, particularly in their impoverished East London neighbourhood. The social scientist was primarily interested in 'citizen-led innovation' and the notions of 'co-design'. The engineer was particularly focused on active community input to design and the creation of prototypes with 'socially accessible' parts and hardware. All were nevertheless agreed that the project would not exist without being facilitated through an arts process. Drawing on my experience to date, I believe this to be due to the fact that, unlike other disciplines, art has no other remit than the construction of meaning, a purpose that guides the totality of the process to achieve its potential. As with any kind of art, my creative role in this project has been to bring together the elements that have presented themselves, adding others as necessary, and then 'holding' the unwieldy alliance until something has begun to take shape, not knowing what would emerge. In this situation the artist is not the driver of the project, but rather the means through which creative ideas become realised.

Nabeel Hamdi (2004) has demonstrated how change starts where one is, and developed from there can rival the sweeping political interventions of those holding political power. The *Active Energy* project was able to take as its starting point the Geezers' direct experience, and through addressing this we arrived at an issue of global significance. Mouffe (2005, 39) has further described how the political erupts in very different places and not only through democratic structures, pointing instead to a series of new resistances that are grassroots-oriented, extra parliamentary and no longer linked to classes or political parties. She claims that these demands have been taking place through a variety of sub-systems on issues that cannot be expressed through traditional political ideologies, and are shaping society from below. I recognise these tendencies as often reflected in two distinct and sometimes overlapping forms in community-based art, as exemplified in the work with the Geezers. The first is that of 'giving voice', one of the key remits of community arts, and articulated by Fitzgerald (2004, 79) as 'the question of power and the right of people to contribute to and participate fully in culture, the right to have a voice and the right to give voice'.[12] Art is an effective means for creating platforms in the public domain where these voices will better be heard, while targeting those who need to listen. Simply being heard can have a transformative effect as noted by Freire (1970, 119), who referred to the inward realisation of his 'educands' of their own inherent power to change both themselves and what is around them. In this sense the articulation and positive acknowledgement of their concerns in the public domain have given members of the Geezers Club increased confidence, which has in turn enabled them to speak in situations such as conferences, university seminars and public events that most would have previously found daunting. Their words and ideas have also appeared in exhibitions, articles and the local press, and they now run their own website, through which they report a range of issues affecting their neighbourhood.

The second main strategy for creating social change through community-based art is the creation of alternative models.[13] In their *Third Text* article of 2008, 'Whither Tactical Media', Gene Ray and Gregory Sholette have highlighted a need for cultural activism to refocus its emphasis by recognising a new social order that is calling for a 'do it yourself' form of tactics. Developing alternatives and demonstrating their effect, as the *Active Energy* project has done, can be a powerful means of shifting social values and perceptions. For example, following the event held to celebrate the installation of the wind turbine on the AgeUK building, members of the housing association responsible for all the sheltered housing in the London Borough of Tower Hamlets invited us to visit them to discuss the integration of renewable energy into the 'new build' schemes. Both these strategies bring the practices, knowledge and skills of local people to a place where the political, social and cultural experience of those least heard in society can enter and affect public discourse.

Fundamental to projects such as this is the building of trust between those involved, achieved through mutual respect and valuing of difference. This lays the foundation for fruitful interaction and the positive working relationships required to sustain the project through both its successes and challenges. This form of art practice is often referred to as 'socially situated', as distinct from the now familiar 'parachuting in' of artists to communities by commissioning bodies with art-world agendas. The creative facilitation employed in *Active Energy* can also be seen as a political act. Drawing out community-held knowledge, then enabling this to interact with other ideas, people and concepts lays the ground for new initiatives to emerge that challenge the status quo by presenting more viable options. As an arts project *Active Energy* cannot of itself bring about change. Its prototype turbines remain open source and the project will not be entering into turbine production. The strength of the arts in this context is to celebrate, express meaning and spark people's imaginations. Just as the content of this project has taken its inspiration from the power of water, the facilitation and support offered through the arts to this small and otherwise powerless group of older men has inspired them to illuminate new ideas and possibilities for others to take up for the future.

Notes

1 This chapter includes extracts from my previous writings in *Art: Process: Change*, Routledge (2017); 'Our Land', *Journal of Heritage Studies* (forthcoming, 2018); 'Engaging Older People in Creative Thinking', in *Oxford Textbook of Creative Arts, Health and Wellbeing*, 2015; and 'Groundswell on the Thames', in *WEAD-Women Environmental Artists Directory* [online] (2014). https://directory.weadar tists.org/category/magazine/issues/7.
2 www.demtech.qmul.ac.uk.
3 The phenomenon of artists working as individuals to make products to sell on the art market is in fact one that stemmed from Western societies and is of less than 500 years' duration. It followed technological and social systems that enabled painting to become portable, then offered for sale through a market economy. Since civilisation began there have nevertheless always been creative producers who have worked in their social groups to express collectively held ideas.
4 SPACE is a visual arts organization based in Hackney, East London that provides creative workspace, residencies, bursaries and training opportunities for artists and has an established reputation for community engagement.
5 http://cspace.org.uk.
6 www.active-energy-london.org.
7 http://lambethfloatingmarsh.org.uk.
8 www.hydrocitizenship.com.
9 www.thames21.org.uk.
10 http://artdotearth.org/green-energy-awards-2016.
11 Information about all these projects can be found at www.cspace.org.uk.
12 One might equally argue that 'giving voice' is somewhat of a misnomer, since communities are often very clear about what they want, it is rather if or how they are listened to that is more the issue.
13 This approach is best exemplified in the *People's Plan for the Royal Docks* (1983) led by the People's Plan Centre in the London Borough of Newham and supported by the Greater London Council's Popular Planning Unit during the campaigning over the future of the London Docklands. This initiative provided a well-researched alternative to the plans for City Airport, addressing the need for homes, jobs and services and driving the proposals for the airport to public inquiry.

References

Fitzgerald, Sandy (ed.) (2004) *An Outburst of Frankness: Community Arts in Ireland – a Reader*. Dublin: Tasc at New Island.

Freire, Paolo. (1970) *Pedagogy of the Oppressed*. London: Penguin.

Hamdi, Nabeel. (2004) *Small Change*. London and Sterling, VA: Earthscan.

Leeson, Loraine. (2014) Groundswell on the Thames: 30 Years of Cultural Activism with London's Riverside Communities. *WEAD-Women Environmental Artists Directory* [online]. https://directory.weadartists.org/category/magazine/issues/7.

Leeson, Loraine. (2015) Engaging Older People in Creative Thinking: The *Active Energy* Project, in Stephen Clift and Paul M. Camic (eds), *Oxford Textbook of Creative Arts, Health and Wellbeing*. Oxford: Oxford University Press (pp. 245–249).

Leeson, Loraine. (2017) *Art: Process: Change – Inside a Socially Situated Practice*. New York and London: Routledge.

Leeson, Loraine. (2018 forthcoming) Our Land: Creative Approaches to the Redevelopment of London's Docklands. *Journal of Heritage Studies*.

Mouffe, Chantal. (2005) *On the Political.* Oxford and New York: Routledge.

Newham Docklands Forum and GLC Popular Planning Unit. 1983. *The People's Plan for the Royal Docks.* London: Newham Docklands Forum and GLC Popular Planning Unit.

Ray, Gene and Gregory Sholette. (2008) Whither Tactical Media. *Third Text* 22(5): 519–524.

Chapter 2

This long river

Luci Gorell Barnes

In early November 2013 I began an inquiry about how stories told to me by my neighbour Jean had contributed to my sense of belonging in the valley where I now live. On 16 November Jean died and my inquiry shifted into a process of grieving. Some of Jean's most vivid memories focused on a nearby stream that is now culverted, and I used creative mapping processes in an attempt to reaffirm my sense of belonging, and combine Jean's memories of the stream with the one flowing in my imagination

5 November

Jean and I sit by the fire. Beyond the half-drawn curtains the morning is grey. The table is piled high with her medication. Smokey Joe dozes under her chair. In the afternoon I watch the ambulance from St Monica's turn the corner and leave. After living in this street forever, Jean has gone. Only a miracle could bring her back, but she is tired and not hoping for one.

12 November

Lin rang this afternoon; they are giving Jean morphine and she is sleeping.

17 November

Yesterday evening Rich and I went to the care home. Jean had died about an hour earlier. I sat and held her hand, then we drove home with the bunch of flowers I had bought her on my lap.

   This Christmas I will not visit her with a tin of shortbread. She will not show me how she is wearing the slip that was Frank's last present to her. I look at the photo of her and me on her 97th birthday and her card to us in shaky handwriting, *with love God Bless you from Jean x.*

18 November

I was planning to write about how listening to her stories had helped me feel part of this valley. Now everything feels fragmented by her death and I am worried that my recently constructed sense of belonging here will unravel without her to anchor it.

I think about meeting her for the first time, about a week after we moved in. I asked if she'd lived here long and she said 'Since 1916'. I said 'Wow, that is long' and she laughed.

Jean told me lots of stories; about magic lantern shows at the Tin Mission, crying on her first day at school because she thought she had been sent to the orphanage, and her sister Nell calling Bob home from the stream. She told me about her aunt who ran the pub and her dad who brought her up after her mother died of Spanish flu. Most often she talked about her beloved Frank, but by the end there was not much talking. I held her hand and offered her sips of water through a straw.

19 November

I keep thinking about Jean telling the story of how her nephew Bob jumped over the stream. I am not even quite sure of where it used to be and then I find it marked on an old map, running along the edge of the allotment. I wonder if that's where Bob played? I feel a bit clearer now – it's always a relief to start working with maps – and I draw one of my own with some of Jean's landmarks and our house on it (Figure 2.1). I feel tentative about drawing in the water though; it has been so well hidden.

20 November

There were lights on at Jean's, and Rich and I went in. Lin and the others were there and things were in piles on chairs. They'd put her slip with the clothes they had chosen for her to be buried in. They found a calendar from 1968, the year Frank died, which they will put in her coffin.

21 November

Jean said there is a spring in the pub garden but it's been paved over for years. I cannot find it on the map but it is clearly marked in my imagination. I have been looking at lots of old maps and there are traces of water everywhere. I wonder if that's why our front wall is so damp?

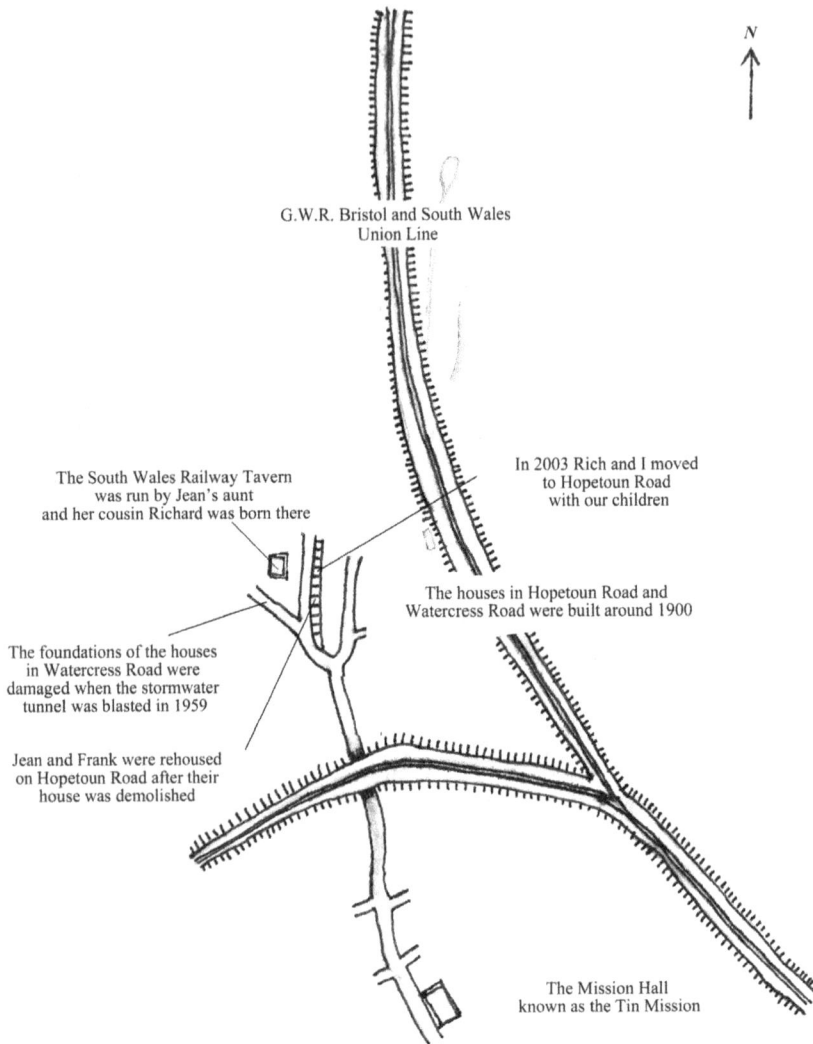

Figure 2.1 Jean's landmarks. Ink drawing manipulated in Photoshop (author, 2013).

22 November

Lin rang to say the funeral is at the church in Stapleton next Friday. The hearse is bringing Jean to the street. I want to have a host of pink carnations, her favourite flowers, adorning her house.

I look at an OS map and see a stream that starts further up the valley, marked as a drain. It joins the spring on Boiling Wells Lane and flows on through the mill garden. I go there, take photographs, and wonder if this is where Bob played. As I walk home, words go round and round in my head as if I am trying to learn them off by heart.

A boy jumps, kicks off from my bank of tangled grass. Suspended above me, he sparkles in sunlight. A woman calls. The boy lands, skids and runs home through emerald watercress beds.

25 November

Val said all the funeral flowers would be put on Jean's front wall as they always do round here.

I thought the spring was on Doreen's farm on Boiling Wells Lane, but the map says it is on Noel's land further up on the left. The big metal gates are padlocked and I cannot get in. Bob said the water bubbles straight out of the ground and that's why it's called Boiling Wells. I can smell the stream in the storm drain as I walk back down the lane and words pound out with my feet, which I write down when I get home.

The hectic mills now sleep, cocooned in quiet dust.
Rains fall, tides turn, and my borders bulge and break. Unleashed, I creep
    under doors, soak corridor carpets and curl muddy slicks round polished
    table legs. In the night I climb trellised wallpaper, defiling photographs of
    lost soldier boys.
Morning finds me sprawled foul breathed on floors. Unwelcome, I am
    brushed shrieking into gutters.
Ruined rugs dry stiffly on front walls.

26 November

I read about Muller raising money to build the orphanage. I always thought it was rather a forbidding place but he sounds kind, and it looks different to me now.

I walk up Stoney Lane remembering my kids careering down it every day after school. I can see how Jean thought she was being sent to the orphanage, the two places are very close.

Muller House is huge. It stands both sides of Ashley Down Road, converted into City of Bristol College and private housing. I sit on the stone steps in front of the flats and write. I imagine I can hear Ali among the voices of children coming from Fairfield School on the other side of the valley, but of course I can't. Stapleton church steeple rises behind Purdown. A lunchtime bell has rung at Fairfield and all is now quiet.

I walk home down the allotment path. It used to be peaceful but now it is a busy cycle route and fluorescent bodies flash by. The map says the stream flows between the allotment and the railway track. Foxes and badgers have hollowed out their homes in its bushy banks and Doreen's geese graze near the rusting carcass of a once-yellow car. I pass the place where we scratched Dan's name into wet cement, and the bank where Jean picked purple flowers when I took her out in her wheelchair. I stop at our allotment and notice that Muller House is perched right above me. I am surprised because in my mind it was further along the valley. I remember that I have not pruned the raspberry canes yet. I am late this year.

27 November

I walk up the allotment path thinking about the foxes. Jean fed them on her front wall every night. Rich would often chat with her on his way home from gigs. I imagine myself as the wall and words turn in my head. Back at home I make a cup of tea and write them down.

> Sooty darkness brings the foxes. On the dot they trot down the street and hop onto my back. Walk along my spine with dainty feet. Sit straight, as if there were white napkins at their throats and silver cutlery held in their neat paws.
> She opens the door, steps gently into the night. I become the table their food is tendered on. She stands in shadow watching them dine. In the morning their empty dishes are strewn at my feet.
> I have not seen her since she fell. She reached for me but I slipped her grasp. For weeks the hopeful foxes kept calling.

28 November

My memory feels unreliable. I have pictured Jean's stories for so long I cannot tell what is real and what I have made up. I want to get my bearings and so I draw another map combining Jean's landmarks with my own (Figure 2.2).

29 November

Rich and I go to the flower market at dawn, returning home with armfuls of pink carnations. We put them in jam jars on Jean's windowsills and lay bunches on her front wall.

We gather outside her house. Jean's cousin talks about Bob being chief of their Stinging Nettle Gang and it reminds me of us as kids, making nettle soup over smoky fires. Bob gives me a newspaper cutting about the Watercress Road houses. There is a photo of Frank pointing to a huge crack

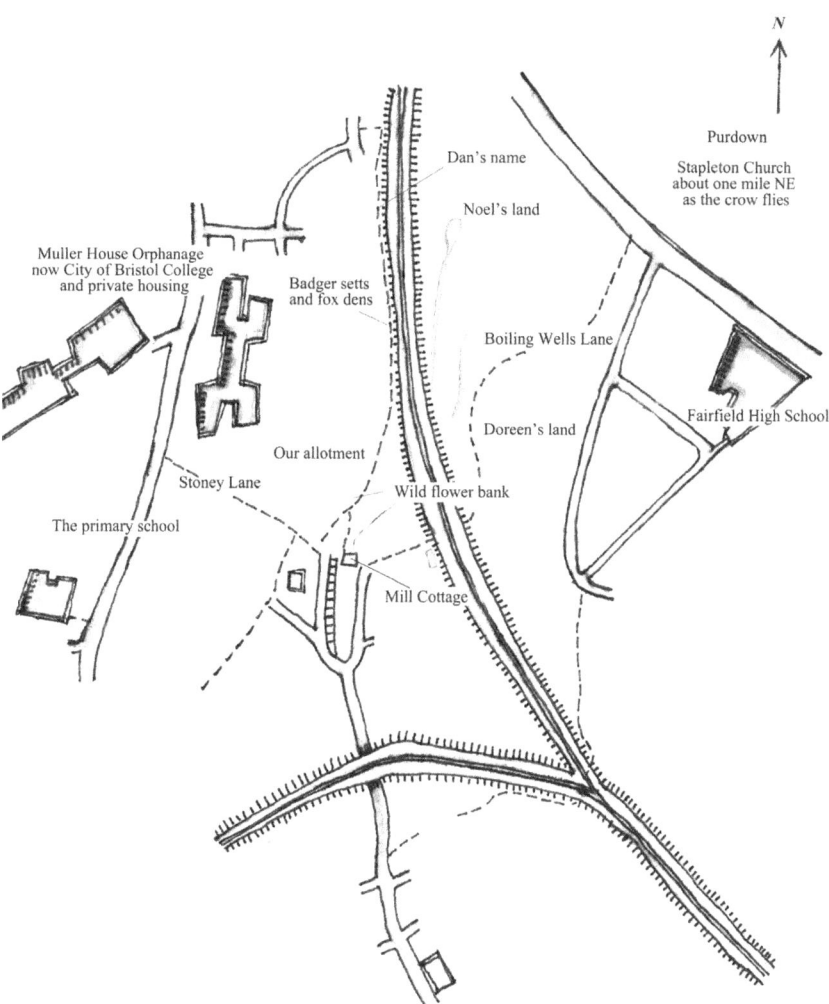

Figure 2.2 Combined landmarks. Ink drawing manipulated in Photoshop (author, 2013).

running down their back wall caused during the construction of the storm-water tunnel (Bennet 1960). Jean is saying 'It would break my heart to leave. This is our home. I am praying for a miracle'.

At the funeral I hear things about Jean's life that she hadn't told me. How she operated a lathe in the war, cycled through the allotments armed with a hatpin in case she was attacked, and that she did not remember her mother, only the black plumed horses and glass sided hearse coming to the street.

After the funeral Rich and I walk beyond Lockleaze, looking vainly for the source of the stream. When we get back the flowers have gone; I guess Lin took them. It feels like Jean's presence is leaving the street.

I have been listening to Billy Bragg's song (Bragg 2013) about the long river having run its course, and it makes me cry every time I hear it.

1 December

I still feel unclear about the course of the stream and go back to the historical maps on the Bristol City Council website.[1] I had not realised how much water shifts: forcing its way up through cracks, before being sent back underground. It seeps and leaks and trying to map it feels like a Sisyphean task.

On the 1880 map a weir marks where the stream meets the spring at Boiling Wells. Together they flow into the millpond, waterfall and watercress beds. On the newer maps the stream has been banished underground. I imagine it crawling blindly through the stormwater tunnel so recklessly blasted with dynamite. Beyond our valley the stream re-surfaces and joins the River Frome near the M32.

2 December

I pick up Jean's stockpile of food for Smokey Joe. Her house feels rather forlorn.

In Tesco I am about to buy a punnet of raspberries as a treat for Jean and then remember . . .

3 December

I draw a map of the ghost streams that run under our houses (Figure 2.3). We have been walking on water all this time.

6 December

The house feels too small and I go outside. The last leaves have fallen leaving summer nests abandoned to the sky. I walk up the allotment path. Ahead of me the sky fades from indigo to green. Dad used to say he liked looking at the stars because it reminded him of his own insignificance. I buy a bar of chocolate and eat it, walking past a column of panting cars. I pass Jean's house and reach my front door. I go in, sit at the kitchen table and write.

> Summer evenings, young neighbours gather. Lean on me with clinking bottles of beer. Bowls of pasta wait for chalky children to finish playing hopscotch. I am annexed by youth as she dozes inside.

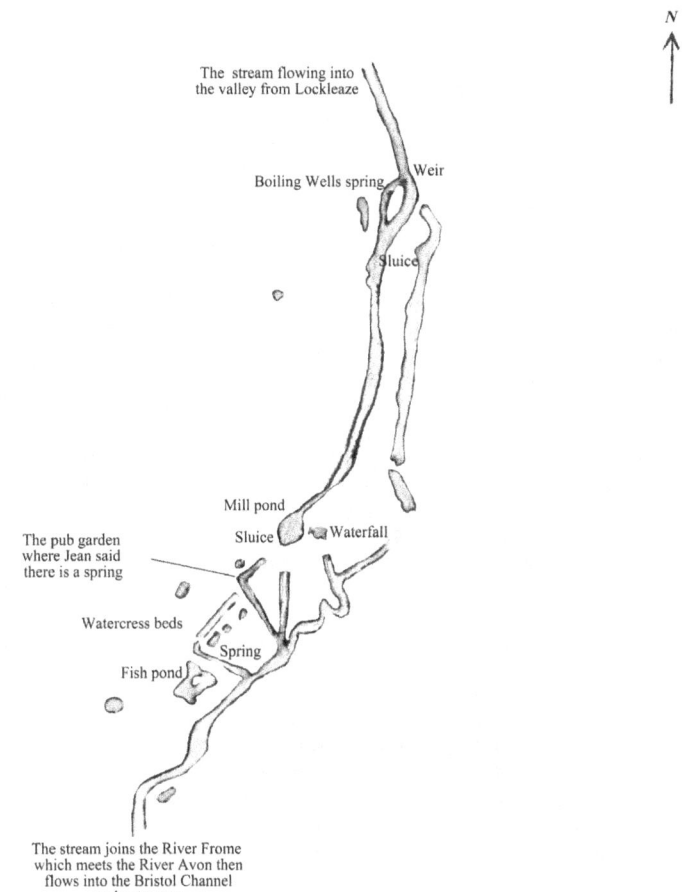

The stream flowing into
the valley from Lockleaze

N

Weir

Boiling Wells spring

Sluice

Mill pond

The pub garden
where Jean said
there is a spring

Sluice   Waterfall

Watercress beds

Spring

Fish pond

The stream joins the River Frome
which meets the River Avon then
flows into the Bristol Channel
and out to sea

Figure 2.3 Ghost rivers. Ink drawing manipulated in Photoshop (author, 2013).

Others hurry in and out, feeding the cat. Olga from Latvia comes to care
three times a day. The summer months roll by in this slow way.
Autumn brings a sudden flurry, and then quiet. The cat is tempted from the
empty house with treats.
They come in black, lay flowers upon my back. Place jars of pink
carnations on the windowsills and weep. The hearse turns in the street.

Reflection

I will now examine Jean's storytelling as a social action through which she constructed her family history, connecting us both to the valley in which we lived. This feels like an appropriate form of analysis in the light of what I know about Jean, but I offer it simply as the way in which I tried to understand 'meanings of the other' (Drewery 2005, 308) rather than claiming any truth or authority for it.

I gave my own voice to her stories, choosing fragments of them that echoed strongly for me, particularly in the weeks following her death. Chase writes about stories as being 'embedded in the interaction between researcher and narrator' (2005, 659), and for me Jean's stories were rooted in our friendship and my sense of home.

In considering the specific ethical issues of this inquiry I resonate with Speedy (2008, 39) who writes:

> I do not have any glib answers to these concerns about representation and the balance of voices, other than to persistently interrogate and seek out my own and other constructions of ethical know-how and constantly and tentatively navigate the landscapes between competing sets of ethical and aesthetic principles.

Jean's niece, Lin, cared for her in a way that enabled Jean to live independently until she died, and I wanted to write something that she felt comfortable with. She agreed to read it and I waited anxiously for her response, feeling I was walking a very delicate line by writing something so soon after Jean's death. I was relieved to hear back from her, and this is part of a longer email in which she also reminisced about 'AJ' her name for her Auntie Jean:

> It did make my cry a little, but it is so wonderful to know that AJ was loved and treasured by your family – she always said you were all part of our extended family and I know that she loved and treasured you all in return.

Although I was not able to discuss my inquiry with Jean I held her in mind throughout, wondering what she would have made of it. I tried to capture the complexities of how I experienced our relationship and hope Jean would have felt my sincere intention to honour our friendship.

Telling stories to create family history

Le Guin argues that we 'force the world to be coherent' (2004, 264) by creating narratives that turn random happenings into structures with plots, reason and purpose. Telling stories is a way of being human, of trying to make sense of circumstances and events by allocating significance to them: 'The meaning of life cannot be comprehended outside the narrative process' (Järvinen 2009, 329).

In narrative inquiry we pay attention to the context in which stories are told in order to interpret the meanings people construct from their lived experience. Geertz (1973, 2) calls this 'thick description', a process in which nuanced portrayals combining description and explanation expound the social significance of what has been observed: 'We need, therefore, to analyse narratives and life materials, in order to treat them as instances of social action – as speech-acts or events with common properties, recurrent structures, cultural conventions and recognisable genres' (Atkinson 2005, 6).

Noticing how someone narrates their life we become aware of the 'versions of self, reality, and experience' (Chase 2005, 657) their story telling produces. This allows us to see something of how they conceive of their place in the world, and better understand them as socially inscribed beings.

I moved in next-door-but-one to Jean in 2003 and there was affection between us from the start. Jean's family had lived in the valley for many years and this is where she spent her whole life. Jean was the only family member still living here but she evoked her relatives' presence by telling stories about them. In the early days of our friendship she told me these stories as we leaned on her front wall. Over time our friendship deepened and I listened to them in her sitting room, against the mounting sound of the television.

Jean's stories seemed to provide a sense of continuity and she narrated what had been regularly occurring events: Bob playing in the stream, having her tea in the pub, and Frank's gifts to her. The repetitive and everyday nature of her stories suggested 'master narratives' (Tannen 2008, 206) of family loyalty, enduring love and physical connection with home:

> That is, narrative is one strategy for ordering the lived experiences, meanings, and sensibilities of a small group into a cultural form that can be understood and passed along to succeeding generations as it is told and retold over time.
>
> (Peterson and Langellier 2006, 179)

By telling her stories Jean seemed to construct meaning and values from her family's culture into a form that could be understood, inherited and passed on.

Creating stories in landscape

Jean's stories took place in this valley. It feels rural here nestled among allotments and farms, and its enclosed nature seems to foster a strong sense of community. Her memories were outlined by the physical boundaries of this place and as a result the landscape itself was part of her narrative: 'Narrative action as we have seen is not set against a backdrop of space and time coordinates, it is crucially constituted in space and time' (Baynham 2003, 365).

Jean told stories from the perspective of this valley as *home* and they evoked a sense of her as being part of the place. Her narratives not only created an identity for her, their reiteration also contributed to my sense of knowing the valley and therefore belonging in it, affecting me beyond what she might have imagined: 'The world of the audience exceeds the grasp of the storyteller' (Peterson and Langellier 2006, 175–176).

Jean and I were neighbours for 11 years, sharing what Rishbeth and Powell (2013, 163), describe as 'mnemonic memories'. We watched the sun touch our garden walls on winter mornings, climbed the steep hill to school, and listened to the stream as it rushed through the storm drain after rain. Most people have a wish to belong, to 'find identity in landscape and place' (Taylor 2008, 2). Our surroundings take on particular meanings through our repeated interactions within them, however to really feel at home a person needs not only to know but also to feel 'intimately known' (Ingold 2000, 403). Our relationship with landscape goes beyond our individual experiences. It is our shared understandings of a place that allow us to belong there, united with the 'coherent identity" (Said 2000, 179) within it: 'This is why landscape and memory are inseparable because landscape is the nerve centre of our personal and collective memories' (Taylor 2008, 4).

We interpret our habitats through common ideas, values and meanings, constructing them as places where people and memories reside.

The cultural knowledge that Jean inscribed upon this valley was part of what made it legible to me, written through her stories: 'People do not move through an abstract biophysical matrix, but through meaningful cultural landscapes, within socially variable envelopes' (Whitridge 2004, 243).

The past lives on in our landscapes, and in my daily life I experienced the 'hollow places' (De Certeau 1988, 108) in which Jean's family slumbered. I agree with Freeman (1998, 46) who writes that current interest in narrative and relational thinking indicates a desire to be part of something larger than our selves. Jean's stories connected me to this place and when she died I felt my sense of relationship here was shaken.

Walking in memories

I had an overwhelming desire to be outside, as if my emotions were too big to be articulated indoors. I followed my impulses, walking every day through the landscape that Jean and I had shared, and my inquiry became an act of remembrance. Specific mourning periods exist across a range of cultures: time set aside in which to grieve and adjust to loss, and I moved into a process of 'maintaining connection and relationship' (Moules et al. 2004, 7) by gathering up the threads that linked me with Jean.

I walked old paths and new, taking short cuts, detours and diversions. The valley bears marks of many different lives upon it, not only physical sites like the mills that have come and gone, but also narrative traces where the landscape is written by events that have taken place in it. I studied maps and researched local sites and stories. I 'read time backwards' (Watson 2008, 334) to try to make sense of the present, approaching Jean's stories in a new, proactive way. I asked questions as I walked, for example wondering who had tended the watercress beds and who paved over the spring. 'The initiative in narrated history does not belong to the past but to the questions asked about it, and these questions are always posed from a specific perspective, the perspective of the present' (Järvinen 2009, 322).

I pictured myself as a filament woven in among others. I thought about unraveling, unpicking and re-stitching myself back into the fabric of the landscape. I felt that Jean's stories were a thread that she had pulled from her past and embroidered for me in the present: 'The understanding of narrative as a making is evident in investigations into the elements, aspects, and structures that make up narrative. Such efforts locate narrative as an object, work, or text that is imagined, fashioned, and formed' (Peterson and Langellier 2006, 174).

It seemed that I was mending the rip that Jean's death had left behind, using stitches that allowed me to see where the edges of our lives had touched.

I kept thinking about Borremans' painting of a rural scene that hovers somewhere between being a landscape and map (Figure 2.4). Huge figures

Figure 2.4 Trickland, oil on canvas (100 × 180 cm), (I-Large) (Michaël Borremans, 2002).

tend the earth as if it were a grave or a garden. Powerful in terms of size, they work obediently, each one an unremarkable part of the group's quiet purpose. I felt myself to be like one of Borreman's figures or Sebald's fishermen who replace each other time and time again 'further than memory can reach' (Sebald 2002, 52).

Every day I walked up muddy tracks with the image of Bob jumping over the stream vivid in my mind, and I became preoccupied with somehow reconciling Jean's memories of the river with the one flowing in my imagination. I returned home to pore over old maps, located the long-gone Ashley Vale Mill, and notated them with my own landmarks. As I stepped back out into the December rain it seemed that maybe the membranes between eras were thinner than I had supposed. Perhaps by combining Jean's and my landmarks through my making and remaking of maps, I was time travelling into what Freeman (1998, 47) calls 'spirals of remembrance and return'. In particular the stream held this sense for me, simultaneously present and absent, and seeming to point beyond human time and anything I could record.

With these thoughts and feelings, I walked as a 'process of *appropriation*' (De Certeau 1988, 97) hoping to reaffirm my sense of belonging. As I walked, words went round my head in time with my muddy footsteps, expressed as if I was the riverbank, or Jean's wall, and then as if I was the river itself. My desire to belong, to be accepted and known in a specific place, expanded into a much wider concept as I experienced a powerful physical and emotional connection to the landscape, and I recalled Siddhartha who said that, 'The river is everywhere at the same time' (Hesse 1922/1954, 83).

Constructing a layered text

I walked and wept, I wrote and read. I notated old maps and drew new ones in an effort to align stories and people. I needed to create something readable from the tangle sprawled across my journal and laptop, and felt the most transparent way to do that was to show my process in the structure of what I wrote. I felt relieved at this idea of representing the different threads as related yet distinct elements, rather than artificially fusing my lines of thought (Figure 2.5).

There was something about the idea of layers keeping the 'research text in motion' (Rath 2012, 443) that I was drawn to. I understood it to mean that my text did not have to be fixed, but could show the reflexive nature of my process. I stopped thinking about amalgamating my ideas and moved into a writing process that shifted between describing, locating and reflecting on them: 'Specifically, becoming engaged in the reflective process helps researchers to develop a strong bond with the writing. It is in this process of reflection-on-action that one is exposed to his/her inner thoughts and feelings' (Johnson-Leslie 2009, 253).

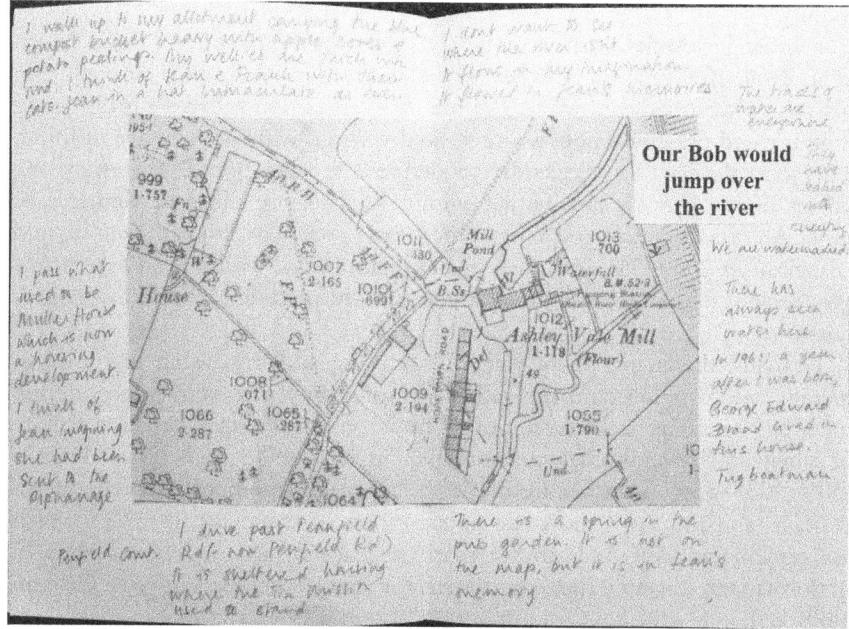

Figure 2.5 Pages from artist's workbook (author, 2013).

In relation to each other the different threads of my inquiry had unlocked moments of understanding for me, and by staying faithful to these moments my layered text emerged.

Ochs and Capps (1996, 31) write: 'The boundaries between selves and other entities are porous'. Certainly, Jean's and my stories bled into each other, like the water seeping through this valley. I represented this by recounting events as if witnessed and uttered by the landscape itself. Imagination is pinned to the physical world but not limited by it, and my approach allowed me to portray my emotions, ideas and interpretations alongside descriptions of events that had taken place:

> The strange idea that reality has an idiom in which it prefers to be described, that its very nature demands we talk about it without fuss – a spade is a spade, a rose is a rose – on pain of illusion, trumpery, and self-bewitchment, leads on to the even stranger idea that if literalism is lost, so is fact.
>
> (Geertz 1988, 140)

I wrote from the intersection between embodiment, memory and fantasy. From this location I gained a sense of events happening over a much longer period than Jean's or my lifetime, drawing me deep into the heart of my inquiry.

I wrote in journal form, a mode that insinuates a 'contract' (Tullis Owen et al. 2009, 189) of truth between writer and reader. This text is derived from my artist's workbook, however the original version is a reflective tool for my own use and presented here is a reworked version, with an audience in mind. The journal text uses a calendar to define a specific period suggesting the phases of my process but nesting within its dates are different elements of time. Autobiographical and biographical time collide as 'human time' is told (Brockmeier 2000, 54). Imagination and memories sit within lived experience that renders chronological time simultaneously pertinent and of no consequence at all.

Drawing the maps was a way of capturing and organising my process, helping me to understand the valley from my own perspective and offering me a way to navigate the terrain of my inquiry. On the maps time is a collage of labels and symbols that position lives in relation to each other. Denzin (2006, 333–334) describes history as a 'montage, moments quoted out of context' and my maps show the juxtaposition of fragments, many of which are no longer here: 'It is thus a mark in place of acts, a relic in place of performances: it is only their remainder, the sign of their erasure' (De Certeau 1988, 35).

Like any writing, map making is process of construction: 'Geographical maps are abstract and concrete at the same time; for all the objectivity of their measurements, they cannot represent reality, merely one interpretation of it' (Schalansky 2009, 10).

I made field notes, remembered stories and consulted other people's maps in order to create my own. Whitridge (2004, 241) writes about the emotional impact of connecting our imaginings with physical places, and I noticed how soothed I felt when I aligned my vision of Bob jumping with the geographical source of the stream. For me, Jean's death had fragmented the landscape and through the process of mapping I reconstructed it for myself: 'The coded visual language of maps is one we all know, but in making maps of our worlds we each have our own dialect' (Harmon 2004, 11). My maps were an intrinsic part of my process, an atlas of stories and grief.

Conclusion

Rath (2009, 150) describes assembling a 'structure of possibilities' in which the reader is invited to fill the spaces between layers with her own interpretations. It is possible that ideas and meanings can be found in the open weave and interplay between my texts and I offer it here as a 'work-in-the-making' (Lather 2004, 5).

The stream embodied a sense of time and place that moved me closer to understanding how we belong in − or maybe belong to − the landscape.

The process of trying to map the stream characterised this inquiry for me. I struggled to establish its shifting course and this felt like a metaphor for the narrative process, something that is always in motion and cannot be fixed. Geertz (1973, 15) describes cultural analysis as being 'intrinsically incomplete' and I have no sense of having arrived at any conclusions. This writing is a tentative version of a lived experience and through it I hope to show how I negotiated meanings working as a 'reflexive practitioner' (Rath 2009, 150).

Observing her writing about her late grandfather Holman-Jones states: 'It did not produce "findings"; it was not generalizable outside of asking audience members to recall and reinhabit their own moments of grieving. It generated whatever credibility it earned out of fumbling attempts to make sense of my loss' (Holman-Jones 2005, 772). Similarly, I have attempted to tell the story of a time of mourning and I offer this account as a means for readers to think 'creatively and imaginatively *with*' me (Geertz 1973, 12). Fairclough describes her grieving process on the nearby Severn estuary and writes: 'my own attempt to throw my grief into the river, the tide carrying breath/voice to the river's source before it is pulled back out to sea' (Jones and Fairclough 2016, 106).

Maybe a similar purpose drew me to my little stream, to wash away my loss. Meanwhile I continue to walk, to remember and imagine. I am still journeying in Jean's footprints; for me they will always be visible in this valley.

Note

1 http://maps.bristol.gov.uk/knowyourplace (accessed 26 November 2013).

References

Atkinson, P. (2005) Qualitative Research: Unity and Diversity, *Forum: Qualitative Social Research*, 3(26). www.qualitative-research.net/index.php/fqs/article/view/4/9 (accessed 27 November 2013).

Baynham, M. (2003) Narratives in Space and Time: Beyond 'Backdrop' Accounts of Narrative Orientation, *Narrative Inquiry*, 13(2), 347–366.

Bennett, R. (1960) *Bristol Evening Post* (January).

Bragg, B. (2013) Goodbye, Goodbye, in *Tooth and Nail* (CD), New York: Cooking Vinyl.

Brockmeier, J. (2000) Autobiographical Time, *Narrative Inquiry*, 10(1),51–73.

Chase, S. (2005) Narrative Inquiry: Multiple Lenses, Approaches, Voices, in Denzin, N. and Lincoln, Y. (eds), *The Sage Handbook of Qualitative Research*, Thousand Oaks, CA: Sage, 651–680.

De Certeau, M. (1988). *The Practice of Everyday Life*, Berkeley, CA: University of California Press.

Denzin, N. (2006) Pedagogy, Performance, and Autoethnography, *Text and Performance Quarterly*, 26(4),333–338.

Drewery, W. (2005) Why We Should Watch What We Say: Position Calls, Everyday Speech and the Production of Relational Subjectivity, *Theory & Psychology*, 15 (3),305–324.

Freeman, M. (1998) Mythical Time, Historical Time and the Narrative Fabric of the Self, *Narrative Inquiry*, 8(1),27–50.

Geertz, C. (1973) Thick Description: Toward an Interpretive Theory of Culture, in *The Interpretation of Cultures: Selected Essays*, New York: Basic Books, 3–30.

Geertz, C. (1988) Being There, in *Works and Lives: The Anthropologist as Author*. Cambridge: Polity Press, 129–149.

Harmon, K. (2004) *You Are Here: Personal Geographies and Other Maps of the Imagination*, New York: Princeton Architectural Press.

Hesse, H. (1922/1954) *Siddhartha*, Owen, P. Trans., London: Penguin.

Holman-Jones, S. (2005) Auto-Ethnography: Making the Personal Political. in Denzin, N. and Lincoln, Y (eds), *Sage Handbook of Qualitative Research*, Thousand Oaks: Sage.

Ingold, T. (2000) *The Perception of the Environment: Essays on Livelihood, Dwelling and Skill*. http://web.zone.ee/bateson/Ingold/The%20Perception%20of%20the%20Environment.doc (accessed 20 December 2013).

Järvinen, M. (2009) Life Histories and the Perspective of the Present, in Harrison, B. (ed.), *Life Story Research, Volume* I, London: Sage, 319–339.

Johnson-Leslie, N. (2009) Taming the 'Beast': The Dance of Sustaining Reflective Practice on the Dissertation Process, *Reflective Practice: International and Multidisciplinary Perspectives*, 10(2), 245–258.

Jones, O. and Fairclough, L. (2016) Sounding Grief: The Severn Estuary as an Emotional Soundscape, *Emotion, Space and Society*, 20, 98–110.

Lather, P. (2004) Getting Lost: Feminist Efforts toward a Double(D) Science, in *Shifting Imaginaries in Curriculum Research: Feminist Poststructural Modes Toward Enhancing Visibility and Credibility*, AERA, San Diego Symposium. http://people.ehe.osu.edu/plather/files/2008/09/gettinglost.pdf (accessed 27 December 2013).

Le Guin, U. (2004) *The Wave in the Mind: Talks and Essays on the Writer, the Reader, and the Imagination*, Boston: Shambhala.

Moules, N., Simonson, K., Prins, M., Angus, P. and Bell, J. (2004) Making Room for Grief: Walking Backwards and Living Forward, *Nursing Inquiry*, 11(2), 99–107.

Ochs, E. and Capps, L. (1996) Narrating the Self, *Annual Review of Anthropology*, 25, 19–43.

Peterson, E. and Langellier, K. (2006) The Performance Turn in Narrative Studies, *Narrative Inquiry*, 16(1), 173–180.

Rath, J. (2009) Writing My Migrant Selves: Using Mystory to Script a Multi-Reflective Account of Context Appropriate Pedagogy, *Reflective Practice: International and Multidisciplinary Perspectives*, 10(2), 149–159.

Rath, J. (2012) Autoethnographic Layering: Recollections, Family Tales, and Dreams, *Qualitative Inquiry*, 18(5), 442–448.

Rishbeth, C. and Powell, M. (2013) Place Attachment and Memory: Landscapes of Belonging as Experienced Post-Migration, *Landscape Research*, 38(2), 160–178.

Said, E. (2000) Invention, Memory, and Place, *Critical Inquiry*, 26(2), 175–192.

Schalansky, J. (2009) *Atlas of Remote Islands*, London: Penguin.

Sebald, W. (2002) *The Rings of Saturn*, London: Vintage.

Speedy, J. (2008) Reflexivities, Liminalities and other Relationships with 'the Space between Us', in Speedy, J., *Narrative Inquiry and Psychotherapy*, Basingstoke: Palgrave Macmillan, 27–43.

Tannen, D. (2008) 'We've Never Been Close, We're Very Different': Three Narrative Types in Sister Discourse, *Narrative Inquiry*, 18(2), 206–229.

Taylor, K. (2008). Landscape and Memory, in *the Right to Landscape: Contesting Landscape and Human Rights*, UNESCO International Workshop www.unesco.org/new/fileadmin/MULTIMEDIA/HQ/CI/CI/pdf/mow/mow_3rd_international_conference_ken_taylor_en.pdf (accessed 28 November 2013).

Tullis Owen, J., McRae, C., Adams, T. and Vitale, A. (2009) Truth Troubles, *Qualitative Inquiry*, 15(178), 178–200.

Watson, C. (2008) Tensions and Aporias in the Narrative Construction of Lives, *Qualitative Research*, 8(3), 333–337.

Whitridge, P. (2004) Landscapes, Houses, Bodies, Things: 'Place' and the Archaeology of Inuit Imaginaries, *Journal of Archaeological Method and Theory*, 11(2), 213–250.

Chapter 3

Sunless waters of forgetfulness (a geopoetic assemblage)

Antony Lyons

Reflections on geopoetics

Before diving into deeper, at times subterranean, waters, I first offer some reflections on the fusion-term *geopoetics*, a creative framing that I've increasingly embraced within the fluid borderlands of natural sciences, landscape studies and anecdotal observation. My borrowing of the term has grown out of long-term involvements with relational 'deep-mapping' processes and 'slow art residency' situations. For me, geopoetics is about elegantly aligning the linear, dendritic forms of rational knowledge creation (scientific, field-data, etc.) with those governed more by the rhizomic[2] imagination (speculation, intuition, lateral thinking, etc.). Defiantly occupying this hybrid terrain, Jacquetta Hawkes' book *A Land* put down a marker for a shift in perspective and consciousness. In it, she speaks of the history of the Earth's crust as having a rhythm, and she imagines 'a unity with trilobites'. This meditative work forms part of the enduring human efforts to braid earth-knowledge with social constructions, and that 'can only be shown as a blurred reflection through hints coming from many directions but always falling short of their objective' (Hawkes, 1951, preface).

I see Hawkes' 'blurred reflection' as an apt metaphor for the inter-relationships between the triad of physical geographical knowledge, cultural enquiry and psycho-poetic intuition. Geopoetics might then be described as a process by which rationality and science are reflected in the mirrors of subjective, imaginative and landscape-situated knowledges. More specifically, in positioning Hawkes' work as one fountainhead, geopoetics forms a conduit for an ever-evolving geo-aesthetic, deep-time awareness (or 'geological turn') that is increasingly in evidence across the spectra of creative and cultural fields. The growing fascination with the 'Anthropocene' epoch, as an environmental, social and artistic meme, is symptomatic of this.[3] A seminal reference to geopoetics as an expanded, fusion concept was made in the early 1960s by the American geologist Harry Hess, in an effort to encourage creative thinking among fellow geologists. This was in the context of the search for the evidential basis of the theory of Plate Tectonics.[4] Since the 1970s, it has become associated with the expanded poetry and writings of Kenneth White,[5] who wrote that 'Geopoetics is concerned, fundamentally, with a relationship to the earth and with the opening of a world' (White 2004, 243).

I am also influenced and inspired by the words of Canadian poet Don McKay, who, in his essay 'Ediacaran and Anthropocene: Poetry As A Reader of Deep Time', speaks of geopoetry as providing 'a crossing point, a bridge over the infamous gulf separating scientific from poetic frames of mind, a gulf which has not served us well, nor the planet we inhabit with so little reverence or grace' (McKay, 2011, 10).

The roots of my hybrid creative enquiries and production grow from a history of study and work within the geosciences and water sciences – realms where a practitioner is frequently called upon to integrate temporal concepts, field evidence, dynamic spatial imaginaries, and reconcile multiple narratives. Perhaps the geopoetic impulses lie deeper still, but the many formative years of study – of landscape encounters, fossickings and fumblings – still reverberate. The tactile and haptic processes of geological fieldwork in the hill and mountain terrains of Ireland and Wales have had an enduring effect. The 'geo-philosopher' Robert Frodeman, in his book *Geo-Logic* (2003), suggests that diverse research settings might benefit from embracing a geo-synthesis and geo-field-based approach. He positions geologic understanding as a 'hermeneutic process' (Frodeman, 1995, 963). Describing a more disruptive, generative liminality, Hyde (2017) refers to the traditional tricksters (Coyote, Hermes, Loki, Eshu, etc.) as

> the sacred boundary crossers ... bringing to the surface the fact that your point of view is always particular, and that you never know enough to read the world. The tricksters are technically what we might call 'hermeneutes'; characters who teach you how to read the world.

The trickster is an interlocutor and questioner, and reminds us when we've sold our souls. In the contemporary creative sphere, it is common to encounter the role of artist-as-mediator, artist-as-bridge (e.g. between the rational and the expressive imaginations), even artist-as-shaman or trickster – much of this harking back to the practice of the artist, Joseph Beuys, in the second half of the twentieth century. In tandem, there is a growing future-oriented movement to occupy the liminal zones between science, society, ecology and art. The Leonardo[6] network is an influential globally operating example of such groundbreaking hybridity.

Faced with hugely complex socio-eco-geo dynamics, Anthropocene concerns and what certain researchers now term 'wicked problems', there is perhaps some comfort to be found in simplification – but this can be a false comfort. In any extended creative investigation of place, one encounters a variety of flows and tensions. These include flows of people, political shifts and upwellings, plus of course ecological and water dynamics.

Donna Haraway (2016, 1) writes:

> Our task is to make trouble, to stir up potent response to devastating events, as well as to settle troubled waters and rebuild quiet places ...

> Staying with the trouble does not require such a relationship to times called the future. In fact, staying with the trouble requires learning to be truly present, not as a vanishing pivot between awful or edenic pasts and apocalyptic or salvific futures, but as mortal critters[7] entwined in myriad unfinished configurations of places, times, matters, meanings.

Water legacies: the eco-psycho-symbolic

The subject here is the exposure, and trajectory, of landscape – and water-scape – relationalities through the lens of water. In the realms of water-consciousness, there are depths that lie largely inaccessible, unless approached through avenues of intuition, imagination and poetics. On the one hand, there is, as Illich pointed out, H_2O (the hygienic and hydraulic waters of utility and industry) – 'a cleaning fluid' – and often privately owned at that. Humans have grown distant from 'the waters of forgetfulness', characterised by the mythic, symbolic, magical and subconscious. 'H_2O and water have become opposites: H_2O, is a social creation of modern times, a resource that is scarce and calls for technical management' (Illich, 1985, 76). Linking this shift to changing attitudes towards excretion and body-odour, Illich challenges the 'modernist' paradigm, and suggests the need to re-embrace the mystery and mythic along with the utility of water. His 'Waters of Forgetfulness' are an echo of the Ancient Greek river Lethe, the 'river of forgetfulness', where souls of the newly deceased left their memories, as they crossed into Hades. He suggests that we still carry deeper symbolic and ritual associations, suppressed though they may be. Perhaps forgetting is not the answer, but a kind of 'unlearn-ing' and 'letting go' may be needed to achieve a confluence of the streams of materiality, empiricism, intuition and dreams? 'Dreaming beside the river, I gave my imagination to the water, the green, clear water, the water that makes the meadows green ... The stream doesn't have to be ours; the water doesn't have to be ours' (Bachelard 1983/1999, 8).

A more biophysically inflected anchor-point in this context is the work of Theodor Schwenk, founder of the Institute of Flow Sciences in 1960, whose influential book *Sensitive Chaos* (1996) builds on the rich tradition of Goethe, Blake, Steiner and others in ascribing subtle meta-physical qualities to the dynamism of water. Echoing Illich, according to Schwenk (1996, 10), people no longer look 'at the being of water but merely at its physical value ... A way of thinking that is directed solely to what is profitable cannot perceive the vital coherence of all things in nature'. Schwenk binds together hydrodynamics (at many scales), biological morphology, metabolism, geology and much more. His work remains unsurpassed as a holistic treatise on water as life-giver and universal essence.

In nature writing, and in art history, we can encounter the familiar, often romanticised, 'water bodies' (rivers, lakes, pools, etc.). However, a different approach, embracing the thinking of 'dark ecology' leads one further, wider, deeper – into plutonic hydrothermal flows, abyssal vents, stratospheric clouds, tsunamis, industrial/sewage effluents – all of which are less readily conversed with, but each as 'real' as a pastoral or sublime riverine landscape. Emerging from a dream-state, the poet, Coleridge (1816), penned the enigmatic *Kubla Khan*:

> Where Alph, the sacred river, ran
> Through caverns measureless to man
> Down to a sunless sea.

In 2007, together with poet Ralph Hoyte, I followed in the footsteps of Coleridge and the Wordsworths in tracing the course of the Holford Brook from its source to the coast.[8] We sought, very much as Coleridge described, a stream 'traced from its source in the hills among the yellow-red moss and conical glass-shaped tufts of bent, to the first break or fall, where its drops become audible, and it begins to form a channel' (Holmes, 1989, 161).

Revisiting my online blog description of that walk:

> Holford Brook Expedition, 8th Oct. 2007: In a Stalker-esque manner, The Poet and The Artist-Scientist[9] converge for the task of following the Holford Brook from source to sea. Having located it, we are unable or unwilling to leave the vicinity of the bubbling source. Intuitively, I think there is an elusive awareness that this point (in space and time) marks an important initiation, or portal, into the strange ways of this land. In the shadow of ancient Dowsborough Hill Fort, it feels to me as if these waters are themselves utterly transformed by their emergence into the light of day, from an unimaginable journey of seeping and coursing through the stony bedrock deep beneath the hills. Groundwater becomes stream-water ... very soon to become seawater.
>
>   Water has shaped these hills; from the wearing down of some 'original' igneous mass into the granular-rocks that now make up the bulk of the sedimentary sequence. In the interim, these crystalline grains may have resided in other layers in other long-denuded hills.

There is a palpable sense of this 'cycling' – not just of the geological materials, but also the water itself, round and round; if not here, then somewhere.

> There is everything here, the sea, woods wild as fancy ever painted, and William and I in a wander by ourselves found out a sequestered waterfall in a dell formed by steep hills covered with full grown timber trees.
>
> (Dorothy Wordsworth)[10]

Coleridge derived a concept he called 'esemplastic'. Along with exploring the difference between 'imagination' and 'fancy', he coined this new word to describe an aspect of the imagination. Based on Greek words meaning 'into' and 'one' and 'mould,' it describes the taking of images, words and feelings from a number of realms of human endeavour and thought, bringing them together through the 'esemplastic power of the poetic imagination'. 'Imagination', for Coleridge, was 'a middle state of mind', 'hovering between images'. If the mind 'is fixed on one image, it becomes understanding; but while it is unfixed and wavering between them, attaching itself permanently to none, it is imagination'. These ideas have been traced to the writings of Jacob Boehme, (b. 1575). It was the (re)discovery of Boehme by the German Romantics, and by Blake and Coleridge in England, which initiated a resurgence of imagination in relation to reason.

In the contemporary creative realm, one powerful example of such interwoven efforts is Chris Marker's 1982 film *Sunless/Sans Soleil* (perhaps even a *Kubla Khan* reference in this title?). In his engrossing assemblage work of visual and verbal poetics, there is still room for a good dose of political and social comment. It is likely that that Coleridge would have applauded. In this film (and others of his), Marker pays homage to the work of Russian director Andrei Tarkovsky. A scene (from Tarkovsky's *Stalker*) that I greatly admire is a three-minute continuous take: a slow meditation, filmed with the camera moving over a shallow pond, at the bottom of which lie scattered a plethora of objects half covered with water-weeds and slime. The scene is exquisitely rich in atmosphere, affect and metaphor – further enhanced by the accompanying soundtrack. For me it strongly recalls memories of observing votive offerings in Irish holy wells and healing springs.

Figure 3.1 is an image from near Lough Gur in County Limerick, Ireland – a watery place to which I've been intimately connected since my early childhood. I recently sought out the nearby St Munchin's Holy Well. The enduring reverence for this well – as evidenced by its continued upkeep and recent votive offerings – to me reflects a strong heritage of deeper pre-Christian sacred attachment, and enchantment, with water. In legend, the lake is said to demand

Figure 3.1 St Munchin's Holy Well, near Lough Gur, County Limerick, Ireland (photo A. Lyons).

a human heart every seven years (Figure 3.2), and the 'Tree of the World', like the Yggdrasil of Norse mythology, is said to lie somewhere in its depths.

More prosaically, there was, historically, a modest eel fishery at the lake. In our time, eels, which were once an important food item, have become a creature that frequently elicits revulsion. The slimy, the muddy, the watery dark is repulsed or suppressed. Here, Jungian insights meet Illich, meet Morton's 'dark ecology'. Dams and habitat destruction (by draining marsh-lands) not only represent domination of flowing, life-enhancing water; they also act to decimate aquatic populations, especially of migratory fish. On the one hand, the current dire situation of the eel is of eco-psycho-symbolic interest; in another way, it is a tale of human exceptionalism, hegemony and erasure of species and life-support systems. I will circle back to this topic later, but diving deeper into sunless eel-dom, we also find 'Sunless Sea',[11] a recent computer game, partly inspired by Coleridge's words. The eel-revulsion is present in the game's text: 'it's an ugly thing! Even uglier dead than alive. Friendlier, though.' Like Ballard's *Drowned World* vision of London beneath the marshy reeds, this popular game presents a fantasy of a London that has fallen beneath the surface of the earth and into a subterranean sea. Underneath, in the

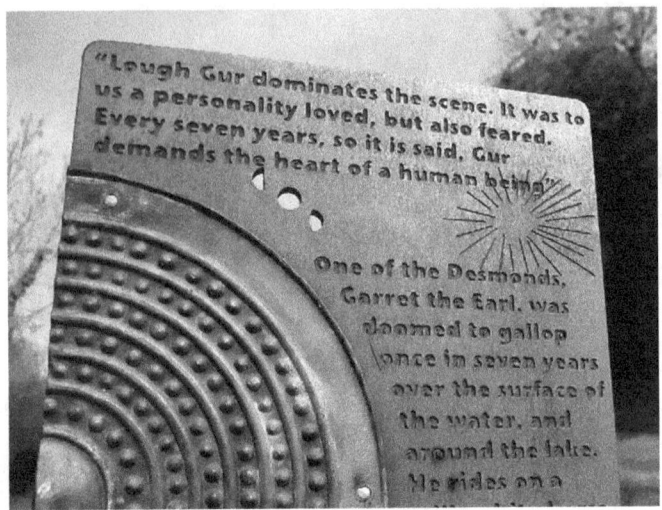

Figure 3.2 Detail from sculpture work *Lough Gur Nexus*, by Antony Lyons/Deiseal 2007 (photo A. Lyons).

darkness, flows a second internal ocean known as the *Unterzee*. In a different nuance to the forgetfulness of Coleridge, Illich and the Greek's River Lethe:

> Death is not the end in Sunless Sea. You are able to pass on scraps of your life to the next comer, an inheritance of your choosing: the chart you so carefully drew (perhaps the most valuable of all assets), your knowledge, cunning, strength or skill. With the passing of each new life there is grief but also opportunity.
>
> (Parkin 2015)

I have also explored eco-symbolic perspectives during extended artist-residency involvements in rural Portugal. To assemble an installation work (*Sacred Water Ways*, 2013), I borrowed a communal 'wash house' or laundry in the small town of Sul. In 12 large water-filled stone sinks were floated 6 different plant materials, with multiple layers of associated meaning, derived from encounters in the landscape and from conversations with local people. The locals' sense of ownership of the immersive installation space – also their own utilitarian space – translated into their 'appropriation' of some of the material contents. They gradually removed bunches of the hundreds of carnation flowers that I'd chosen to incorporate because of locally strong religious and political overtones. One association is that the flower is the symbol of the 1974 'carnation revolution', the *coup d'etat* that ended Salazar's 'Estado Novo' regime. From the wash house,

Figure 3.3 Part of the installation space for *Sacred Water Ways*, Portugal, 2013 (photo A. Lyons).

the flowers were gradually relocated to the nearby homes and small religious shrines.

Given the inherent ambiguities and unboundedness of a geopoetic practice, the works readily slide towards the metaphysical realms, appreciating the porosity of human–nature–spirit boundaries. Glimmers of long-vanished shamanistic or animistic ways may emerge via the revisiting of myths and stories, such as in a very early incantation-like composition called *The Song of Amairgen*, which I incorporated as a live recital piece in the *Sacred Water Ways* performative installation (Figure 3.3). In this text, there is an imaginative fusion, or identification, with both living and non-living aspects of the environment.

I am the wind on the sea;
I am the wave of the sea;
I am the bull of seven battles;
I am the eagle on the rock;
I am a flash from the sun;

I am the most beautiful of plants;
I am a strong wild boar;
I am a salmon in the water;
I am a lake in the plain.[12]

As well as being an expression of Illich's duality of water (practical, quotidian laundry/ablution, fused with the symbolic), the central role, in this installation space, of a poetic dream-like cinematic work also facilitated a further expansiveness of encounter, especially with watery affect. As Benjamin (1935/1969, 235) pointed out (speaking of film, and art's presence in time and space): 'By close-ups of the things around us, by focusing on hidden details of familiar objects, by exploring common place milieus under the ingenious guidance of the camera, . . .[it] introduces us to unconscious optics as does psychoanalysis to unconscious impulses'.

Water, through its material presence, sonic qualities and more mysterious resonances provided the matrix for a form of 'communitas' (Turner, 1969). As environmental artist, Reiko Goto (2013) writes, 'Touching water is a beginning of the discourse to listen to others, including people, things and the environment at deeper level'. Our perceptions and sensibilities are often locked into our routine sensory and time relationships to the world; we are trapped by our senses and cognition. Other living/ dynamic systems reveal extraordinary, marvellous, uncanny attributes when perceived on a microscopic level or via shifted time frames. From macro to micro, hidden cycles, rhythms and pulses emerge; vastly more powerful and enduring than our vainglorious and ultimately futile efforts to control. In our era, attempts are being made to forge, or rediscover, new ecological relationship paradigms. However, the powerful forces of the status quo are not to be underestimated. 'Now I warm towards the concept of neighbourliness as a template with which to approach and write about our relations with our fellow organisms. It permits concern, shared circumstance, even love from afar, but demands no reciprocity' (Mabey 2013).

Embodied, sensorial creative research activities overlap with ethnographic fields of study with increasing permeability. Both critical cultural and geographical scholarship have begun to embrace the reality of affect and emotional attachments to place, alongside concerns with the 'more than human' realms. Terry Eagleton (1990, 14) speaks of aesthetics as existing 'between the material and the immaterial: between things and thoughts, sensations and ideas, that which is bound up with our creaturely life as opposed to that which conducts some shadowy existence in the recesses of the mind'. Linked to this is the vibrant materialism expounded by Jane Bennett (2010), who aims to 'think slowly' and challenge 'the idea of matter as passive stuff, as raw, brute, or inert'. She talks of matter having a 'nonhuman vitality', which 'arrives through humans but not entirely because of them' (p. 17). These ideas resonate strongly with me, and through my work, and are central to creative geopoetic ways of encountering landscapes and their component materials.

Rising tides: the local and the global seas

I draw on a background of both scientific and creative involvements with water landscapes, especially around the Severn Estuary in the UK. This strongly tidal zone has witnessed ebbings and flowings of debates and plans to dam, or barrage, the tidal waters. In the discourse are to be found topics that include heritage management (tangible and intangible), energy policy, regional economics, coastal resilience and ecology. This is a watery site where the energies of deep ecology bubble up and come face-to-face with powerful commercial and industrial forces and agendas.

Embracing a transdisciplinary and personal response to this situation, I have adopted a geopoetic, deep-mapping approach. The setting of the Severn Estuary elicits a fusion of long-term human relationships to coastal landscapes, together with the topic of the damming of rivers to create reservoirs for hydropower generation and water supply. Among the proliferation of dam and flood-defence projects worldwide, little attention is given to the complex vitality of watercourses; nor to the voices of those who relate to these water environments in emotional, creative, visceral, 'sentimental'[13] ways; nor indeed to any intrinsic sacred status. The original vast floodplain of the Severn Estuary has, since Roman times, been progressively 'reclaimed' for farmland. The current situation is recognised as precarious and unsustainable; yet there is resistance to even speculating on reconnecting the river to its former overspill zone. Instead of the ecologically boosting relinquishment of land areas, flood-control efforts are largely focused on hard-defences and deep dredging of tributary rivers. In the end, in time, the land areas will eventually be 'reclaimed' by the river and sea. The infusion of creative perspectives can play an important part in rectifying cultural separation and catalysing diverse new responses to change and disruption, such as threats related to sea-level rise. According to artist-facilitator, Simon Read (2017):

> We have immanent on our coast a strong sense of ownership of the coastal landscape and resistance to change but an acceptance of responsibility to intervene in its governance is not so evident. That this needs addressing is certain and from my point of view the arts are strongly implicated in a state of affairs that has come about through the historical separation of communities from the direct sense of responsibility for their own landscapes.

Much of my work is prompted by a sense of disconnect or conflict. In attempting to weave new tapestries of perception, my adopted creative methods incorporate open-ended fieldwork and experimental remixing of archives, recordings, data and contemporary narratives. The film-poem *Transgression – Regression* (2017) weaves disparate time frames (geo-, ice age, tide-, etc.) and is inspired by the definition of 'transgression' as a geological term describing an advance of the sea over land areas. In a poetic and critical vein, the film is a mosaic of speculations on climate change, sea-level rise, ecological crisis, power relationships, governance and resistance. Pervading much of my work, there is a fascination with diverse tidal situations that share some common challenges, such as conservation of intertidal/coastal heritage sites, biodiversity loss, the survival of traditional fishing activities and the looming development of tidal- and/or nuclear-power infrastructure. Through film works and installation assemblages (*Sabrina Dreaming*, 2013; see Figure 3.4), I am exploring the tensions that exist between some enduring 'taskscape' or coastal livelihood activities and the extractivist industrial geo-engineering forces operating in the Severn Estuary landscape. Within sight of familiar, solid ground, the mudflats are very different in character; a place apart – shape-shifting, perilous, uncertain, alien. The subject of 'time' also comes very much to the fore. The sensation of being in the intertidal zone is one of transiting a space that exists 'out of time' or in cyclical rather than linear time. There is a sense of disconnect, a stepping out of the world, and a dissolving of the human (plus a dissolving of subject–object separation) in this setting. It is only in recent years that human communities have become distanced from the visceral reality of this mud-zone. At one time

Figure 3.4 Still from *Sabrina Dreaming*, a work-in-progress film-poem (Antony Lyons).

there existed a host of 'push-net', 'lave-net' and 'putcher' fishermen jostling for position in the low-tide channels. Prior to the mid-1960s the crossing of the estuary was solely by means of small and vulnerable ferryboats. For travellers, boatmen and fishermen, there was deep respect for the powerful sublime forces of tidal currents and deadly fickle weather.

Echoing this theme of watery disconnection, in 2006 I installed a suite of ephemeral sculptural and sonic interventions around the harbour of the nearby city of Bristol, which focused on the manner in which the tide has been physically and psychologically banished from the heart of the city during the 200 years of the industrial era. Coinciding with the emerging global dominance of mechanical 'clock-time' in the organisation and structuring of life and work, the impulse to distance the lunar/tidal presence, with its slow rhythms (and muddiness), has been paralleled by disconnection from, and destruction of, ecosystems, habitats and the biosphere as a whole. This experiential distancing is I believe a factor in enabling the behemoth of hegemonic, extractive modern practices to proceed unabated. Similar patterns and impulses are also to be observed within the history of dam-building and reservoir creation.

We are entering a new era for the role of creativity in public discourse and in social–ecological relationships; creativity developing in the shadow of the converging crises of ecosystem collapse, violent destabilisation of human populations and of our psyches – when techno-industrial science veers towards the dead-end of the anti-human, anti-ecological and anti-life. Today, these sit alongside a renewal of the concept of *biophilia*[14] (as well as *hydrophilia)* and a resurgence of Lovelock's concept of *Gaia*. Adding fuel to the fire of some current debates we find Timothy Morton's *Ecology Without Nature* (2009) in which he challenges the romantic 'otherness' conception of nature as something raised on a pedestal, external to us. Furthermore, in *The Ecological Thought* (Morton, 2010), we encounter what he terms the burgeon-ing 'hyper-objects' and 'wicked problems' of global warming and sea-level rise. Morton suggests that we need to perceive these problems through radically expanded intimacy and thinking – temporally and spatially. We can no longer maintain a Cartesian distance or objectivity. Similarly, in Felix Guattari's highlighting of 'mental ecology' and the pollution of the uncon-scious, he also forged provocative vital connections (Guattari, 1989). The task of exposing the contamination continues, and can be a slow-burn process. In the early 1990s I encountered Bookchin and his 'social ecology' ideas, which focused on the socio-political implications for the environment and ecology, while critiquing 'deep ecology' for its isolationist, socially blind attitudes. Today, I find I am drawn to other radical shifts e.g. to the 'Dark Mountain' discussions,[15] to philosopher-poet Derrick Jensen's writings on 'de-civilisa-tion' (Jensen, 2006) and the expanding rewilding debates. I'm conscious that these too are also very exposed to Bookchin's criticism of falsely separating environmental and the socio-cultural realities.

Watery liminality: 'the sacred river ran ...'

Regarding the shoreline, or intertide, as a liminal place of possibility, contestation and tension, as well as resistance and emancipation, this in-between zone may be described as a 'heterotopian other space' (after Foucault, 1967).

'In civilisations without ships (i.e. heterotopias of some sort) the dreams dry up, espionage takes the place of adventure, and the police that of corsairs'. There is a complex and deep human connection to the coast, the seaside, the beach. I am interested in how this space can be both a liberating life-enhancing bioregion and a 'necro-region' (after Iovino, 2012) where geological processes of destruction are laid bare. Endless dynamism and flux operate on the land-water boundaries and the impermanence of all human efforts is also revealed as if in a mirror. And there is perhaps a certain relief in that?

Our cultural myths contain liminality too. Characters in surviving ancient myths and texts are frequently presented as ambiguous, slippery or unsentimental in the sense of their good or evil intent. There is the aforementioned trickster 'Loki', the goddess 'Kali', or the Irish example of the *Cailleach Bhéara* (hag, or 'old mother' of Beara), who is portrayed as a powerful, ageless goddess ruling over dreams and hidden realities. The cailleach is both a proud, wise woman, and equally a malevolent crone figure. She appears in the Irish *Dindsenchas* ('lore of places'), which has pre-Christian origins. This is a body of work in prose and poetry, along with descriptions, part factual, part fantastical, of the etymologies, topographies and cosmologies of places, as well descriptions of totemic animals. It is an ingredient in the geopoetic, deep-mapping, metaphor-producing meshes with which I work. In a liminal watery context, she is connected with rivers, lakes, wells, marshes, the sea and storms. In Scotland, the last burst of harsh winter weather is called A' Chailleach. She has a role in creating floods and sits near water, washing the clothes of the dead.

Whether oral or written (e.g. on papyrus), human expressiveness and communication has long been tied up with water. Arguably this is evidenced too by ancient moulded-clay figurines, and cave walls adorned with hand-drawn pigment animals. Water is a carrier, an enabler. We are water; we express water; we cannot master water. Unboundedness is the nature of water; likewise of imagination and creative research. Acceptance of ambiguous perspectives – along with inherent paradoxes and chance events – can embrace both insignificance and hope. Don McKay (2011, 38) addresses this when he calls for 'attention to release its grip on fixed principles, to risk radical not-knowing without succumbing to the seductive currents which go by the name of nihilism' and, then echoing Hawkes (1951, 24): 'we become members of deep time, along with trilobites and Ediacaran period organisms ... we give up mastery and gain mutuality'.

Notes

1 Joyce (1939), chapter 1.
2 'Let us summarise the principal characteristics of the rhizome: unlike trees or their roots, the rhizome connects any point to any other point, and its traits are not necessarily linked to trait of the same nature; it brings into play very different regimes of signs, and even non-sign states' (Deleuze and Guattari, 1987, 23).
3 Bruce Clarke (2014) offers up a critique of the wholesale 'bandwagon' adoption of this concept as 'an exercise in rebranding', and one that 'deflects Earth System Science from its Gaian inspiration'.
4 Specifically, the search was for the evidence base of the process of 'sea-floor spreading'. Hess borrowed the term 'geopoetry' from the work of Dutch geophysicist Johannes Umbgrove (1947).
5 *The Scottish Centre for Geopoetics* is one legacy of these endeavours www.geopoetics.org.uk.
6 Leonardo/ISAST today is the leading organisation for artists, scientists and others interested in the application of contemporary science and technology to the arts and music. www.leonardo.info.
7 As Haraway (2016, 169) writes, 'In this book, "critters" refers promiscuously to microbes, plants, animals, humans and nonhumans, and sometimes even to machines'.
8 This durational artist-residency project in Somerset, England was funded by the Quantock Hills AONB, and is documented at www.quantockdreaming.net.
9 In the film *Stalker* (1979, directed by Andrei Tarkovsky) two central 'characters' are 'Writer' and 'Professor'.
10 Letter from Dorothy Wordsworth to an unknown corresdpondent, 4 July 1797 (cited in Wordsworth, 1935). My blog can be found at www.quantockdreaming.net (entry dated 8 October 2007).
11 Produced in 2015 by Fail Better Games, www.failbettergames.com/sunless/.
12 The Song of Amairgen (Amergin), is commonly accepted as the oldest surviving Irish poem. The earliest written text is found in *Lebar na Núachongbála* (*The Book of Leinster*, compiled *c.*1160 AD). This translated segment is by Lady Gregory, in *Gods and Fighting Men: The Story of the Tuatha De Danaan and of the Fianna of Ireland*, arranged and put into English by Lady Augusta Gregory, London: J. Murray, 1904.
13 'Sentimentality originally indicated the reliance on feelings as a guide to truth, but current usage defines it as 'an appeal to shallow, uncomplicated emotions at the expense of reason'. https://en.wikipedia.org/wiki/Sentimentality.
14 Edward O. Wilson's hypothesis that there is an instinctive bond between human beings and other living systems, which was popularised by the Icelandic singer Bjork in a multi-media project in 2011.
15 The Dark Mountain Project is a network of writers, artists and thinkers who have stopped believing the stories our civilisation tells itself. We see that the world is entering an age of ecological collapse, material contraction and social and political unravelling, and we want our cultural responses to reflect this reality rather than denying it. http://dark-mountain.net/about/faqs/.

References

Bachelard, Gaston. (1983/1999) *Eau et les rêves*. English trans. Dallas: Pegasus Foundation, Dallas Institute of Humanities and Culture.

Benjamin, Walter. (1935/1969) The Work of Art in the Age of Mechanical Reproduction. In Hannah Arendt (ed.), *Illuminations*, translated by Harry Zohn. New York: Schocken Books, p. 218.

Bennett, Jane. (2010) *Vibrant Matter: A Political Ecology of Things*. Durham, NC: Duke University Press.

Clarke, Bruce (2014). The Anthropocene, or, Gaia Shrugs. *Journal of Contemporary Archaeology* 1(1), 101–104.

Coleridge, Samuel. (1816) *Christabel; Kubla Khan, a Vision; The Pains of Sleep, Volume 1*. London: John Murray.

Deleuze, G. and Guattari, F. (1987) *A Thousand Plateaus: Capitalism and Schizophrenia*. Minneapolis, MN: University of Minnesota Press.

Eagleton. (1990) *The Ideology of the Aesthetic*. Oxford: Blackwell.

Frodeman, Robert. (1995) Geological Reasoning: Geology as an Interpretive and Historical Science. *Geological Society of America Bulletin*, 107(8), 960–968.

Frodeman, Robert. (2003) *Geo-Logic: Breaking Ground Between Philosophy and the Earth Sciences*. New York: State University of New York Press.

Foucault, Michel. (1967) Of Other Spaces: Utopias and Heterotopias (Des Espace Autres). Text published by the French journal *Architecture/Mouvement/Continuité* in October, 1984, based on a lecture given by Foucault in March 1967.

Goto, Reiko. (2013) More than Human Research: Blog. www.morethanhumanresearch.com (accessed 5 December 2017).

Gregory, Lady Augusta. (1904) *Gods and Fighting Men: The Story of the Tuatha De Danaan and of the Fianna of Ireland*. London: J. Murray.

Guattari, Felix. (1989) *The Three Ecologies*. London: Athlone Press (originally translated into English by Chris Turner, 'Techno-Ecologies'. *New Formations*, 8, 131–147.)

Haraway, Donna. (2016) *Staying with the Trouble, Making Kin in the Chthulucene*. Durham, NC: Duke University Press.

Hawkes, Jacquetta. (1951) *A Land*. Stirling, UK: Crescent Press.

Holmes, Richard. (1989) *Coleridge: Early Visions, 1772–1804*. New York: Pantheon.

Hyde, Lewis. (2017, 8 May) BBC Radio 4 'Start The Week'.

Illich, Ivan. (1985) *H2O and the Waters of Forgetfulness*. Dallas: Dallas Institute of Humanities & Culture.

Iovino, Serenella. (2012) Restoring the Imagination of Place. Narrative Reinhabitation and the Po Valley. In Tom Lynch, Cheryll Glotfelty and Karla Armbruster (eds), *The Bioregional Imagination. Literature, Ecology and Place*. Athens: University of Georgia, pp. 100–117.

Jensen, Derrick. (2006) *Endgame*. New York: Seven Stories Press.

Joyce, James (1939) Finnegans Wake. London: Faber & Faber.

Mabey, Richard. (2013, July 18) In Defence of Nature Writing. *Guardian*, www.theguardian.com/books/2013/jul/18/richard-mabey-defence-nature-writing (accessed 23 January 2018).

McKay, Don. (2011) Ediacaran and Anthropocene: Poetry as a Reader of Deep Time. In *The Shell of the Tortoise*. Kentville, Nova Scotia: Gaspareau Press.

Marker, Chris. (1982) *Sans Soleil* [film]. Argos Films.

Morton, Timothy. (2009) *Ecology Without Nature: Rethinking Environmental Aesthetics*. Cambridge, MA: Harvard University Press.

Morton, Timothy. (2010) *The Ecological Thought*. Cambridge, MA: Harvard University Press.

Parkin, Simon. (2015) Sunless Sea Review. www.eurogamer.net/articles/2015-02-06-sunless-sea-review (accessed 21 December 2017).

Read, Simon. (2017) Art, Resilience and Porosity in the Coastal Zone. Presentation at the Royal Geographical Society annual conference, London, August.

Schwenk, Theodore. (1996) *Sensitive Chaos: Creation of Flowing Forms in Water and Air*. London: Rudolf Steiner Press.

Turner, V. (1969) *The Ritual Process, Structure & Anti-Structure*. Ithaca, NY: Cornell University Press.

Umbgrove, Johannes H. F. (1947) *The Pulse of the Earth*. The Hague: Martinus Nijhoff.

White, Kenneth. (2004) *Geopoetics: Place, Culture, World*. Edinburgh: Alba.

Chapter 4

From Gallura to the Fens

Communities performing stories of water

Lyndsey Bakewell, Antonia Liguori and Michael Wilson

On the evening of 7 June 2016 about 60 people gathered at a small museum of rural life in the town of Ramsey on the western border of the Cambridgeshire Fenlands. Among them were seven storytellers, a musician, a cartoonist, an illustrator, a small group of academic researchers and a couple of caterers. The rest were local people who had come to make up the audience and listen to stories about water use and management in the Fens. During the following three hours they listened to stories from farmers, environmental projects, local historians and governance agencies of how the different interests and priorities played out in respect of water resource management, and then they were invited to respond with their own thoughts and stories. They listened to, and joined in with, a song that had been specially composed for the occasion. They had their pictures drawn and put on display. And they sat down, old friends and strangers together, to enjoy a meal of local produce. As the late theatre director and playwright John McGrath might have said, they had 'a good night out' (McGrath 1981).

This was the first iteration of a performance event called *The Reasons*, a forum for public storytelling, devised by researchers from Loughborough University with the support of the Arts and Humanities Research Council's Connected Communities Programme, as part of the DRY (Drought, Risk and You) Project, one of the initiatives funded by RCUK under the Drought and Water Scarcity Programme.[1] The framework for the event was that of a mock courtroom, complete with a judge to act as a master of ceremonies and a jury, of mixed age and gender, selected from among the audience members. The inspiration for this came from a visit to Sardinia in October 2015 by one of the authors (Wilson) at the invitation of another research initiative, the CADWAGO project (Climate Adaptation and Water Governance), to the final workshop of this internationally collaborative project.

One of the key events of the workshop occurred on 15 October, when the participants were invited to attend an evening billed as *La Rasgioni*, a traditional Sardinian (or to be more precise, Galluran) reconciliation tool, which had been revived and staged specifically for CADWAGO. *La Rasgioni* (trans. *The Reasons*) is a traditional tool for settling disputes within communities and dispensing justice that was used in Gallura until the early 1960s as an alternative to the official court system.[2]

This traditional custom is best illustrated by an oral testimony from November 1963, gathered by the local ethnographer Pietro Sassu (www. archivisassu.org/), in which different groups of participants are described as convening to discuss their 'reasons' and to reach collectively a common agreement, where law and common sense come together in the same discourse (Sassu 2009). All people involved in this 'judicial' procedure are part of the community, since the principle aim of 'La Rasgioni' is to preserve community cohesion and to reinforce local relationships by co-designing the resolution to a conflict:

> Traditionally, *La Rasgioni* was used as a public mediation process for reconciling two parties whenever a conflict, often regarding property or livestock, could not be solved in other ways. Five people were involved in the process of reconciliation: two *Alligadori* (lawyers) who represented the interests of each party; three *rasgiunanti* (judges: two nominated by each side, and the *omu di mezu*, arbiter, chosen by both). Once the *Alligadori* had presented their speeches, the *rasgiunanti* examined relevant documents and afterwards were called to pronounce their final verdict. Two solutions were possible: *dizisa* – if the request of one of the parties was accepted; or *arrangiu* – when a compromise solution was suggested. Everyone could attend the debate, thus turning private conflicts into collective public events in order to create learning opportunities for the whole community.
>
> (Ruiu et al. 2017, 2)

The mediation between conflicting 'reasons' is a key part of this event and is aimed at facilitating the recognition of the verdict as a collectively reached decision, rather than one imposed by a representative of the official institutions of the state. It is this element that deeply distinguishes *La Rasgioni* from another traditional practice, common elsewhere in Sardinia, codified as 'Codice Barbaricino', which is based on a so called 'code of honour' acknowledged by the community, but conducted as 'vendetta'[3] (Pigliaru, 2000). In effect the participation of all the community in the decision-making process in 'La Rasgioni' aims to guarantee that the verdict is both effective and durable.

Alongside its contrasting rocky, mountainous and pastured landsape, Gallura's climate and economic infrastructure present a range of conflicts and dilemmas for the community around water use and management, something that *La Rasgioni* helped to resolve. As Pietro Sassu[4] records, this last oral testimony[5] keeps alive the memory of a practice commonly used to resolve conflicts outside of the official legal system. In its traditional form, *La Rasgioni* was the bringing together of local people to discuss pressing and practical issues. Held in the local square, both sides of the story were heard. *La Rasgioni* was both the decision maker and peace-keeper in Gallura: "'La Rasgioni" aimed not only to solve disputes quickly and peacefully, but primarily to restore the pre-existing relationships negatively affected by the conflict, thus preserving the community cohesion' (Ruiu et al. 2017, 2).

Indeed, the more we learn about the process of this historic technique, the more its distinctiveness becomes clear. For example, the collective ritual that lies at the core of *La Rasgioni* is crucial in understanding the active role of the audience. Furthermore, as part of the ritualized decision-making process, all participants generally convene together to eat; usually before the pronouncement of the verdict. This simple collective act underpins the convivial atmosphere of *La Rasgioni* and improves the spirit of community cohesion, and is an aspect of the event that is greatly valued by all participants as an essential component of the ritual. Perhaps due to the high level of illiteracy in Gallura (Lissia 1904), this verbal form of community-based reconciliation was handed down through the generations, surviving as a useful and important form of conflict resolution until nearly 60 years ago, at which point it fell into disuse.

The event that was staged in Sardinia in 2016, devised by Pier Paolo Roggero along with his colleagues from the University of Sassari and the theatre director Sante Maurizi, set out to explore the validity of the traditional tool for community cohesion; testing whether it is possible to air local conflicts in order to construct a shared vision for the future.

This contemporary iteration utilised the fixed roles and structure from the original in order to protect the unique purpose of this decision-making practice. The *Omu di mezzu* – judge – was played by a real judge. The *Rasgiunanti* – the jury – was represented by 40 researchers from the CAD-WAGO project. This represented a shift from the original form, where the level of education did not play any part in the attribution of 'roles' and 'responsibilities' during the arbitration. Historically, it was respect, trust and a certain level of dialectic ability that were the main criteria in choosing the *rasgiunanti*, qualities that were considered most important in rendering them the most capable of interpreting opinions and reaching a final decision. Finally, the *Alligadori* – the lawyers – were played by the leading academics from CADWAGO, representing the opposing sets of interests in the debate, with institutions (such as local government, water companies, governance organisations) on one side, and farmers, private businesses and environmental movements on the other. For CADWAGO's staging, all the people involved in this 'juridical' procedure, except the *Rasgiunanti* (jury), were part of the community, seeking to strengthen the aim of *La Rasgioni* as a means of preserving community cohesion and reinforcing local relationships.

The traditional form of *La Rasgioni* relies on the traditional and collective trust in *so omines*, the 'wise people' of the village, not necessarily elderly, but experienced in the themes touched by the dispute. These *omines*, so called *rasgiunanti*, are perceived as 'custodians of an equal society', 'whose actions are guided by *reason* and not by emotions' (Sassu 2007, 95). CADWAGO's staging maintained this feature as a means of re-constructing the collective trust of the community. Regardless of any traditional misconceptions about the role of emotions in decision-making, CADWAGO's *La Rasgioni* demonstrated that feelings of empathy and trust, excitement or disappointment, were crucial in driving the dynamics of the storytelling process in both the traditional and the re-imagined forms.[6]

As previously stated, it was this re-imagining of *La Rasgioni* in 2016, which led to the question of whether the event could be further adapted for use in other contexts, specifically whether a version of *La Rasgioni* could be created for a British context and whether this could contribute towards the DRY project's core aim of bringing together scientific and narrative knowledge to work together for more effective drought and water shortage planning.

The Reasons in Ramsey

The Reasons took place on 7 June 2016 at a rural museum in the small village of Ramsey, on the western border of the Fenlands. Ramsey Rural Museum is a restored seventeenth-century farm building that now houses a café, exhibition space, display chemist and cobblers shop, along with a pumping and trades room, all of which trace the history of trade and life in the area. Also held at the museum are surviving farming, pumping, armed forces and archaeological artefacts. As a community resource, the museum offered a combination of both familiarity to the audience, along with a flexibility that allowed us to create an intimate performance space. It was always our intention to place the idea of 'performance' at the centre of the event, albeit within a frame of a mock ritual (we could, of course, not recreate an existing ritual that did not exist within a Fenland context). We were sharing with the community a new way of developing community and solving issues.

The Cambridgeshire Fenlands where Ramsey sits is a unique landscape, flat and largely beneath sea level, whose features have been created through the draining and reclamation of land for agriculture and, specifically, food production. The seemingly mono-dimensional nature of the landscape, however, conceals a complex set of interests and aspirations around water management. Without effective water management and the use of pumps, the landscape would return to its natural watery state, and yet the diversity of stakeholders and their competing needs is today far more complex than it was when the Fens were first drained. The needs of agriculture now compete with the needs of environmental projects, anglers, tourism, industry and householders and the governance institutions such as the water companies, the Environment Agency, the Middle Level Commissioners, etc., all have a role to play in balancing these demands. The close proximity of the city of Peterborough to the Fens, with one of the UK's fastest growing populations, also has a significant impact upon water demand in the area. The rapid expansion of the city is in many ways threatening the Fenlands and its careful balance of water management, resulting in a conflict not only around water needs, but also between urban and rural. These tensions that potentially emerge from these conflicting interests, a complex hierarchy of needs, a range of water management options, and contrasting views for future planning were all key issues we were hoping to address through our staging of *The Reasons*.

To begin with we convened a small group of stakeholders from the area to discuss what the event might look like. We met in a small pub on the edge of the Fens and heard a variety of opinions and a range of knowledge and expertise. We discovered a rough landscape of human perceptions, in which the dominant elements were represented by a profound sense of disconnection between Peterborough and the rural area. There was also a general acknowledgement that there needed to be an addressing of the balance between the natural environment and an artificial system for water management. But overwhelmingly there was a common tendency to project the community towards a future that must be different. In this meeting, we began to see the richness and diversity of voices from farmers, conservationists, utility providers and historians as a means of correcting and changing the wide-ranging issue of water usage in the Fenlands. Through this meeting we were able to begin to identify and recruit our team of storytellers for *The Reasons*.

In terms of the structure for the event, it was decided to maintain the key structural components of *La Rasgioni*, while also drawing upon the traditions of the 'ceilidh', as an occasion where communities would gather to sing, play music and tell stories for mutual entertainment. This meant that *The Reasons* would necessarily acquire a more informal and irreverent tone. While we were keen to maintain the idea of the 'courtroom' as a contextual and performative device, we also wanted to foreground *The Reasons* as first and foremost a social event (Figure 4.1). As such it was decided that *The Reasons* would be framed by a specially commissioned song,[7] as had been the case the previous year in Sardinia. Not only did this announce the beginning and end of the event, but it also helped set the appropriate tone: in Sardinia it established the ritualised aspects of the event; in the Fens it set up the event as an evening primarily of entertainment and conviviality. On both occasions, it also helped provide a strongly local context to the subsequent components of the evening.

The opening song was followed by a welcome address by the judge. This was intended to introduce the format of the evening to the audience, to establish the informal tone of the evening and to set up the theatrical device of the mock courtroom. It was also the role of the judge to appoint the jury at this point. In Sardinia, the restaged version of *La Rasgioni* had used visiting members of the CADWAGO project for the jury. Our original intention had

Figure 4.1 The Reasons at Ramsey Rural Museum.

been to recruit a jury in advance from local schools, thus making the young people of the community the ultimate arbiters of the future. Unfortunately this had not been possible due to the particular demands of the school calendar and the exam timetables, so we decided instead to incorporate 'jury selection' into the performance. Members of the jury were thus selected (press-ganged might be a more accurate description) in comedic fashion by the judge, who adorned each appointee with an identifying pink sash.

It would, at this point, be worth discussing in more detail the role of the judge in *The Reasons*. Whereas in Sardinia *La Rasgioni* was traditionally a genuine juridical process,[8] albeit one that sat outside the official legal channels, for *The Reasons* we wanted to adopt the notion of the courtroom as a device, a theatrical frame, which would allow the proceedings to be 'managed', but that also would introduce the opportunity for subversion, comedy and satire. The judge was, therefore, to be the 'emcee' of the evening, with his own comic patter and with the job of maintaining the conviviality and good humour of the occasion. For this we drew upon two main sources: the European folk figure of the Lord of Misrule (or Mock King, Mock Mayor, etc.), traditionally a member of the community elected to preside over the anarchic and festivities of the Feast of Fools; and the character of the judge, Azdak, from Bertolt Brecht's play *The Caucasian Chalk Circle* (*Der kaukasische Kreidekreis*). The rascal Azdak, the local drunk, is an unwilling judge and seemingly most ill-suited to the role, yet proves himself to be the wisest and most humane of men.

This role of the judge was played by Mick Gowar, a writer and academic from the local university (Anglia Ruskin), who arrived abruptly on to the 'stage' in a costume comprising of an academic gown, a bright pink sash bearing the word 'judge' and a large ladle (in place of a ceremonial mace) for 'the serving of justice and the ladling out of wise words'. Sharron Kraus, a folksinger and songwriter, announced the start of the evening with a song, specially composed for the evening from discussions and interviews with members of the local community over the preceding few weeks.

The Judge then outlined the order of proceedings to the audience and distributed further pink sashes to selected audience members as a way of assigning the roles of jury members. This was a kind of improvised clowning routine that served to break down social barriers and created a strong sense of the social gathering that we were aiming to create with *The Reasons*. The remainder of the first part of the evening belonged to the storytellers.

One by one they were introduced and allowed five minutes to tell their stories (at least this was the guidance given to the storytellers – in reality the judge exercised flexibility). In total there were eight storytellers (Luke Abblitt and Mat Smith, both local farmers; Kate Carver from the Great Fen Project; Allan Mott from the Warboys Local History Society; Paul Hammett from the National Farmers' Union; Stewart Howe from the Fenland Trust; Kye Jerrom from the Environment Agency; and David Thomas from the Middle Level Commissioners). Interestingly each storyteller took a slightly different performative approach to their task. Some spoke 'off the cuff' until their time was up, while others spoke from notes (either from paper or from their phone). Some had memorised their story and others read from a prepared text. Each, according to their confidence and prior experience of public speaking, adopted their own, personalised strategy.

Upon conclusion of the stories there was time for questions from the floor, chaired by the judge. The majority of the questions were polite and benign, although one particular intervention exposed the particular tensions between the needs of the agricultural industry for high-quality arable land for food production and the Great Fen project, which is returning some land to its pre-drainage state to allow the regeneration of the peat bog and an increase in biodiversity and natural habitats. It was upon this note all participants retired to the neighbouring marquee for refreshment.

It would be doing it a disservice to describe the ensuing 45 minutes as an interval. The main activity was the serving of a meal by local caterers, but it was also the period in which the jury would meet with the judge to discuss the stories that had been told and the questions raised by the audience. We also staged a pop-up exhibition of the sketches that had been produced by our courtroom artist (Ellise Wilkinson, a graduate artist from Anglia Ruskin University) and our caricaturist (John Elsom) (Figure 4.2). The main purpose of this extended 'interval', we had assumed would be to create a convivial atmosphere, through the sharing of food and drink, to build a sense of community cohesion. Very soon, however, we realised that it was much more than this. It was, in the first instance, a comment by Paul Hammett of the National Farmers' Union that drew our attention to what had started to happen. 'Have you seen what is happening?' he said. 'People are sitting around, laughing, smiling and having a good time. And they are talking about *water*. I've never seen that before.' We had

Figure 4.2 John Elsom's caricature of the jury delivering its verdict at Ramsey Rural Museum.

assumed that this interlude in the proceedings would be in the service of the main performance event, helping build the sense of audience and providing the jury with deliberation time. In fact, it seemed to be the other way round. It was here, around the trestle tables, where the real discussion and debate was taking place. The detail of these discussions was spontaneous and, inevitably, uncapturable, but what seemed to be happening was further storytelling – people exchanging their own personal experiences of living with (and without) water. The storytelling and other performances in the first part of the evening had simply enabled this to happen. And so it was with a sense of renewed energy and understanding of the issues at stake that the audience returned to the main performance space for the second (now the *third*) part of the event.

The final part of the evening was much shorter in length. The jury was invited to deliver its verdict. It had chosen to elect two co-forepersons, so this was delivered jointly and, in concordance with the underlying purpose of *La Rasgioni* to suggest ways in which the community could design its own inclusive solutions to challenges, rather than declare winners and losers, a summary of the stories was provided, along with an analysis of the key dilemmas facing the community and a clear declaration of intent that only through collective action could solutions be found to these dilemmas and conflicting interests.

The performance finished with a closing speech from the Judge and a final rendition of the song, including a couple of new verses that had been written during the interlude in response to the stories and discussions that had taken place. Two incidences of audience response are worthy of note here. First, in an act of spontaneous community singing, the whole audience, without invitation, joined in with the chorus of the song. Second, even after the event was formally drawn to a close, the audience seemed reluctant to leave, preferring instead to continue the conversations and take another drink. In themselves these may not seem like remarkable moments, but together they seem to indicate the extent to which community and a sense of solidarity and common purpose had been built through the event, as well as a serious level of engagement with an issue that policy makers, governance agencies and stakeholders have found it difficult to motivate the public about, except in times of crisis, such as flood or drought.

The Reasons in Peterborough[9]

As successful as this first iteration of *The Reasons* appears to have been, it had always been our intention to restage the event in urban Peterborough, having made any necessary modifications to the format and the programme itself. Wary of falling into the trap of self-congratulation in the euphoric wake of a successful performance event, we held a series of evaluative discussions among the research team and with stakeholders and one key challenge emerged, namely that we had perhaps placed too much emphasis on building community cohesion and agreement and been too cautious about exposing disagreements and provoking debate. We felt that in the second iteration of *The Reasons* we should be bolder in this respect if we were genuinely to offer opportunities for communities to address their dilemmas in a constructive way.

To this end we turned to Augusto Boal's 'Forum Theatre' and specifically the figure of the 'Joker' and the 'Joker System'. Forum Theatre, as conceived by the Brazilian theatre director-turned-politician, is a form by which theatre is used as a collaborative tool to enable communities to explore issues of social and political importance. Formerly passive audiences become proactive spectators (or 'spectactors', as Boal calls them), intervening in and shaping the drama along with the actors. It is theatre as an exercise in democracy, or what Boal called 'a rehearsal for revolution' (Boal 1979). At its core is the Joker System, whose core principle, according to Ruth Laurian Bowman is 'to upset or destabilize the singular reality of the world as it is represented ... in order to explore alternate ways of representing and interpreting that world' (1997, 139–140).

The Joker (named after the neutral card in the pack) plays a particular role within Boal's theatre, namely that of an onstage director or facilitator (or 'difficultator'), whose purpose is not to take sides, but to intervene, challenge, provoke and generally help the momentum of the discussions and debates flow. While our original intention had been that this role would have been taken on solely by the judge, we concluded that the judge was too embedded within the action (and particularly in the management of the storytellers and the jury) to be able to fully take on all aspects of the role, so we created an additional character, the provocateur, whose job was to act as inquisitor and ask the questions that the audience might be unwilling to ask. The role was taken on by Sean Lang, another colleague from Anglia Ruskin University. He would sit in the audience, suitably attired in his bright waistcoat and bow tie, and act on their behalf, with a licence to disrupt and misbehave, if necessary. With his assistance, members of the audience were able to express their feelings more freely and the narratives became collective as shared moments were recognised and built up into communal experiences. During the interval, he would transform into the 'foreman of the jury', helping them to reach their verdict by questioning, challenging and even contradicting each thought and opinion until a collectively agreed conclusion had been reached. Thus he was also able to provide a link between the various elements of the performance. In this way, the figure of Boal's joker was represented by both the judge *and* the provocateur in 'The Reasons'.

A further innovation in the second iteration of *The Reasons* in Peterborough was a partnership with Metal, a local charity that works with creative artists and that was holding a month-long festival with a focus on food production, waste and water. The festival was taking place in the City

Museum, once the site of the City's hospital, and we were able to use this as a venue for the event itself. The performance venue, with its large gallery spaces, was a stark contrast to roughcast intimacy of the rural museum in Ramsey, and an exhibition of the work of artists Lucy and Jorge, addressing issues of community food production and food waste, which the audience was encouraged to visit, provided an appropriate backdrop to the performance.

In terms of format, the Peterborough event followed along the same lines as that in Ramsey: it began with a song, the arrival of the judge and the appointment of the Jury. About half of our storytellers were veterans from the Ramsey event, while the other half was performing in *The Reasons* for the first time. The room was full and the audience much more diverse than it had been in Ramsey. After the stories had been told and audience questions fielded (with the help of the provocateur), everyone retired for an extended intermission for a communal meal. Once again this proved to be not a break in the performance, but its very core. The meal was a symbol of community, of sharing in the local identity. It also acted as a way of further eliciting discussion, and providing personal pockets of time to listen to more experiences and opinions. The meal, both in the staging in Ramsey and in Peterborough, was part of the performance itself, but in Peterborough it took on new meaning. Here the meal was a celebration of a locality many of the spectators were unaware of. It provided the opportunity for the chef to showcase her own talents and the quality of the local produce. In effect, the serving and sharing of the food became a symbolic act of community-building.

After the meal, the sated audience returned to hear the 'verdict', this time delivered on behalf of the jury by the provocateur. As at Ramsey, it was not so much a judgement, but more a plea for collective action and respect for diverse interests. At its core was a call for sustainability, couched in terms of concern about the environmental legacy being created for future generations. The event was finally brought to its ritualistic conclusion by way of a reprise of the opening song and a formal (almost ceremonial) closing of the proceedings by the judge.

Evaluating *The Reasons*

Following the second staging of *The Reasons*, we asked the storytellers to come together one final time to share food with us, and to reflect on their experience of taking part. Discussion around the table quickly turned to the effectiveness of storytelling as a method for sharing problems and reaching conclusions. One storyteller commended *The Reasons* as a new way of engaging people as they had learned 'a great deal about the complexities of water management', which they might not have known otherwise, a reflection that was also commonly expressed on audience feedback forms. On this level *The Reasons* exceeded our expectations: not only were we able to adapt the traditional Sardinian form to appeal to those in the Fenlands and Peterborough, but the events had met both educational and social purposes. Reflections on the meal-as-performance were also positive and identified the continuation of knowledge sharing as a key feature. As one storyteller observed:

> It brought everyone together, literally and figuratively, in communication after what could have been a divisive part of the evening. It also provided a light-hearted and relaxed central point to the evening. It also helped to emphasise the 'show' nature of the exchange of ideas – as opposed to arguments and discussions on social media.

Adopting the Sardinian method of conflict resolution was also seen to be positive. One storyteller found 'the interaction between the Fens and Sardinia [. . .] an unexpected cultural interaction'. We had invited Sardinian observers to both events and they had engaged with audience and performers to discuss similarities and differences with *La Rasgioni*. Likewise, a spectator remarked that they were 'fascinated to hear about this as a historic mechanism for conflict resolution', questioning whether or not this kind of project might do more for the exchange of cultural traditions and knowledges.

Finally, the storytellers were keen to understand if it might be possible to adapt the form even further. They were intrigued by what other areas of interest and decision-making needed to be addressed: 'we need to try work-ing the format in different scenarios, different sectors and making partnerships with policy makers and other stakeholders. Also, we need to ensure the visibility of this work.'

Closing thoughts

Interpreting and re-adapting the original form of *La Rasgioni* as a forum for public storytelling in *The Reasons* represents the intentional choice to prioritise the story-listening process over the conflict resolution function of the Sardinian form. With the community in the Fens, we co-designed a new performative practice to enable unheard voices to be listened to, with the hope of reaching those who are generally disengaged from environmental issues. Throughout the process we were guided by the idea of storytelling (and specifically story-telling as a social activity and live performance event) as a tool to trigger discussions on themes that were under-debated and as a methodology to enhance social learning. Local stakeholders recognised the fluid dimension of stories and their ability to enable people to travel both back in time and into the future and to connect self and other. And this general and immediate under-standing of storytelling as a sort of 'holistic thinking' (Meadows and Kidd 2009) was for them also a way of challenging current social (often hierarchical) dynamics that had been established through more formal contexts, and to bring about a more communal and interactive way of discussing environmental issues that took account of the key challenges they had faced.

In particular, language was one of the main barriers to communication at community level, especially with regard to the impenetrable way in which policy makers and scientists often communicate. In this instance, *The Reasons* was able to mediate the barriers by drawing on personal experience to inform and question the scientific hierarchy of knowledge. Storytelling was able to democratise the public debate and facilitate a broader understanding of the key dilemmas. By doing so we equipped the storytellers and community members with an alternative way of making decisions that could be applied anywhere, to any given topic.

In turn, we sought to develop 'polyphonic narratives' (Marino 2013) where no voices are dominant over other voices, because they are all relevant and they have all 'equal status'. Through a structured but consciously un-controlled creative process, the storytelling-storylistening circle was 'activated' to include multiple voices and identify shared responsibilities for future decision-making. While this process may have uncovered conflicts and dilemmas, it also discovered unexpected common ground in the dialogue between lay and expert narratives, due to the authenticity of personal stories and the natural 'mess' of the world that storytelling both exposes and helps us navigate through uncertainty (Wilson 2014).

Furthermore, the sharing of stories and memories from different sectors of society during the same event revealed a critical tension around the interaction between opinions and facts and stimulated a deeper reflection on the beliefs that drive people's behaviours. Stories can be a way of beginning a conversation that allows for a better engagement with different truths. If 'misconceptions' are perceived to complicate the picture, they also expand the horizon of the debate: challengeable narratives 'provoke' discussions, enable us to elicit counter-narra-tives and bring different stories together, while challenging the notion that there can only be a single truth about a given situation. And this was crucial in *The Reasons*, as we opened a forum for public storytelling to bring together different types of knowledge, facilitate social learning and help shape 'the bricolage of the here' (Leyshon and Bull 2011). Having opened that forum, the challenge inevitably remains, going forward, how to keep it open in a way that contributes to more engaged policy-making.

Notes

1 For video documentation of the first iteration of *The Reasons* at Ramsey Rural Museum, please see the following link: 'There's Something in the Water', The Reasons – Stories About Water Usage, Drought and the Future of the Fens, Ramsey Rural Museum, 7 June 2016, www.youtube.com/watch?v=hjBO4cNUOJE.
2 There are many other examples of performance rituals and traditions being used as a contemporary tool of conflict resolution. See, for example, Ananda Breed's study of how the traditional Gacaca Court and ritual dance were employed to address the legacy of genocide in post-conflict Rwanda (Breed 2014).

3 In English the term 'vendetta' has become inextricably linked with the murderous reprisal of the Sicilian Mafia. Here it refers to any emotional and disproportionate reaction that escalates a dispute. It is the very opposite of reason. Also, where *La Rasgioni* focuses on the community (collective), 'vendetta' is more likely to focus on the individual. 'Vendetta' is reaction, rather than response.
4 www.archivisassu.org/.
5 The voice recordings have been stored at the 'Discoteca di Stato', www.icbsa.it/ Ref: AELM 106/1; 303673 and AELM 106/2; 303674.
6 For further discussion about the role of emotions in everyday life and decision-making concerning social and environmental policy, see Anderson and Smith (2001).
7 'A River Is a Snake', by Sharron Kraus, www.youtube.com/watch? v=1tAgtQkafJc.
8 The American psychologist Jerome Bruner (2003) offers an insightful discussion around storytelling and the role of orality within a Western legal system that privileges written testimony and document over oral and traditional knowledge.
9 For video documentation of the second iteration of *The Reasons* at Peterborough Museum and Art Gallery, please see the following link: 'Think Water': Storytelling for the Future of Peterborough and the Fens, 1 December 2016, Peterborough Museum and Art Gallery, www.youtube.com/watch?v=c4l1KlDZaek.

References

Anderson, K. and Smith, S. (2001) Emotional Geographies. *Transactions of the Institute of British Geographers*, 26, 7–10.
Boal, A. (1979) *Theatre of the Oppressed*. London: Pluto Press.
Bowman, R.L. (1997) 'Joking' with the Classics: Using Boal's Joker System in the Performance Classroom. *Theatre Topics*, 7(2), 139–151.
Breed, A. (2014) *Performing the Nation: Genocide, Justice, Reconciliation*. Kolkata: Seagull Books.
Bruner, J. (2003) *Making Stories: Law, Literature, Life*. Cambridge, MA: Harvard University Press.
Leyshon, M. and Bull, J. (2011) The Bricolage of the Here: Young People's Narratives of Identity in the Countryside. *Social & Cultural Geography*, 12(2), 159–180.
Lissia, S. (1904) *La Gallura*. La Maddalena: Tempio.
McGrath, J. (1981) *A Good Night Out: Popular Theatre – Audience, Class and Form*. London: Methuen.
Marino, M.C. (2013) The Mobile Story: Narrative Practices with Locative Technologies. In Farman, J. (ed.), *The Mobile Story: Narrative Practices with Mobile Technologies*. New York: Routledge (pp. 290–304).
Meadows, D. and Kidd, J. (2009) Capture Wales, the BBC Digital Storytelling Project. In Hartley, J. and McWilliam, K. (eds), *Story Circle: Digital Storytelling Around the World*. Oxford: Wiley-Blackwell (pp. 91–117).
Pigliaru, A. (2000) *Il banditismo in Sardegna. La vendetta barbaricina*. Nuoro: Il Maestrale.
Ruiu, M.L., Maurizi, S., Sassu, S., Seddaiu, G., Zuin, O., Blackmore, C. and Paolo Roggero, P. (2017) Re-Staging La Rasgioni: Lessons Learned from Transforming a Traditional Form of Conflict Resolution to Engage Stakeholders in Agricultural Water Governance. *Water*, 9, 297. doi:10.3390/w9040297.
Sassu, S. (2007) Ordinamenti giuridici di tradizione orale in Sardegna: il caso della rasgioni gallurese. *Sociologia del Dirittto*, 2, 85–112.

Sassu, S. (2009) *La Rasgioni In Gallura. La risoluzione dei conflitti nella cultura degli Stazzi.* Roma: Armando Editore.
Wilson, M. (2014) 'Another Fine Mess': The Condition of Storytelling in the Digital Age. *Narrative Culture*, 1(2), 125–144.

Other resources

http://ramseyruralmuseum.co.uk.
Kraus, S. 'A River Is a Snake'. www.youtube.com/watch?v=1tAgtQkafJc.
'There's Something in the Water': The Reasons – Stories About Water Usage, Drought and the Future of the Fens, Ramsey Rural Museum, 7 June 2016. www.youtube.com/watch?v=hjBO4cNUOJE.
'Think Water': Storytelling for the Future of Peterborough and the Fens, 1 December 2016, Peterborough Museum and Art Gallery. www.youtube.com/watch?v=c4l1KlDZaek.
www.archivisassu.org/.
www.icbsa.it/Ref: AELM 106/1; 303673 and AELM 106/2; 303674.

Part II

Becoming water bodies

Chapter 5

Mapping a blue trace
An intermittent swimming life

Ronan Foley

Introduction: swimming as a relational practice

For people who swim, their relationships with water are powerful and affecting. Many narratives, especially those relating to health and wellness, attribute this to regular practice in specific indoor and outdoor locations (Parr, 2011; Sherr, 2012). Yet many swimmers experience a much more uneven yet no less meaningful set of experiences with water across their life course. This chapter takes an (auto-) ethnographic approach to mapping a blue trace through personal and recounted life-course stories of swimming, in mostly outdoor swimming locations. Though primarily based on Irish accounts, the chapter describes encounters at a range of scales and periods from a range of informants, including the author's own intermittent swimming life. Those resonant encounters show how body/environment are sustained within family histories and cultural practices that include long gaps, bursts of activity and an interplay with wider relational geographies of water. At the heart of the narrative is a concern to uncover the ways in which swimming can be mapped across lives and spaces, with a specific focus on the affective and accretive threads that help to produce health and well-being.

While sport and leisure geographies focus on physiology and elite practitioners (Bolton and Martin, 2013), for many everyday swimmers, its always ongoing role as a non-competitive and sustained practice through their lives is what matters. Personal health and wellness are a big part of their stories, expressed both physically and emotionally through regular and often structured interactions with and within blue space (Throsby, 2013; Bell et al. 2015). Such narratives will inform this piece, but the chapter additionally argues that there is an *intermittent* practice that differs somewhat from the everyday. For intermittent swimmers, the affective and accretive power of irregular exposures can be no less important than regular ones. In particular, an intermittent embodied consciousness can develop, as bodies and bodily relations with places change over time, from childhood through to late-life swimming. One of the unique elements of swimming is that can remain an embodied practice across the entire life course, but is always relational to one's own place in the world, within families and cultures and across geographies and life events. This applies just as much to the places in which we swim; in turn relationally connected to imagination, memory, company and a shared inter-subjectivity.

The focus of this chapter will be an (auto-) ethnography of a global and relational geographical practice running almost as a blue trace across the life course; both mine and others. As a very irregular swimmer who grew up in the interior of Ireland, that blue vein drew me to the sea and continues to draw me back. As a child, I was taken each summer to different coastal holiday resorts, mostly on Ireland's west coast, where the initial dye (*sic*) was cast, and the process of affective layering began. Through much of my adult life, as a migrant in the UK and through widening travels, swimming took a back seat. But the thread was occasionally revived on holidays in warm places with warm water, a valued contrast to the cold waters I grew up around. Since my return to Ireland in 2003, I have rediscovered the seas as both a site of study and play. I now live a ten-minute walk from the sea, yet it feels that everyday there is something new that deepens, redraws or tightens that connective thread, augmented by every single swim, story or wave. The blue trace is always linked to family and friends, especially my father, and through him, many other fathers, mothers, partners, brothers, sisters, friends and often perfect strangers.

Many of these connections coalesce in specific shared public settings – intermittent blue spaces – that continue to be, to a greater or lesser extent, open to all ages, sizes and shapes. In emphasising the value of such spaces, place and personal memory have a powerful affective bond (Seamon, 2013). In tracing an intermittent practice that by its nature requires one to reconnect to one's own body and mind, the affective value of swimming is always accompanied by an embodied reality. For health geographers, such therapeutic settings have enormous potential for physical and mental health treatments, maintenance and recovery. This chapter takes a tentative step in articulating some of the ways these emerge across meaningful if uneven relational experiences with water.

Writing on swimming and blue space

Swimming is having a *zeitgeist* moment. It feels as if every time one opens a page these days, or switches on a screen, there is a new piece about water, the coast or the sea, often featuring swimming at the heart of the narrative. This has been driven in part by the popularity of 'wild swimming', especially in the UK (Rew, 2009), but also in non-fiction writing by what have been described as 'new topographers' (Deakin, 1999; Macfarlane, 2010; Robinson, 2012). For such writers, a close engagement with nature in all its forms uses an assemblage approach that blends together oral history, family memory and place lore with active engagement in nature through walking, sailing and especially in the case of Roger Deakin, swimming. His book *Waterlog* has been particularly influential, documenting a personal blue trace from swimming in his own Suffolk moat, to freezing lakes in Snowdonia and river pools in Somerset. In addition, there has been an explicit documentation of outdoor swimming places using a range of new media, which has led to a developing industry around places that have been simultaneously hidden in plain sight and inaccessible, an arguably double-edged process (Rew, 2009; MacEvilly and O'Reilly, 2016).

More specifically, the close relationships between lived lives, memory and landscape are increasing tropes for social and cultural geographers. From Hayden Lorimer's exploration of a Scottish seaside, micro-geography and John Wylie's exploration of absence and presence on a coastal Cornish walk, to Owain Jones tidal geographies, there are intriguing glimpses of themes central to this chapter (Wylie, 2005; Jones, 2011; Lorimer, 2015). For both Lorimer and Wylie, there was an implicit concern for memory and mood and the potential of the sea to provide an affective trigger that becomes marked in place by specific memorial forms (pet cemeteries, benches) and also by specific embodied practices (sifting, sitting, immersion). For Jones, the very particular permeable nature of tidal coastal waters acts as a valuable metaphor for swimmers, who have similarly permeable relationships with tides and their own bodies, in and out of water.

Recent work by health geographers on the health-enabling potential of nature is increasingly framed within studies of green and blue spaces (Ward-Thompson, 2011; Mitchell, 2013; Foley and Kistemann, 2015) but also through historical research on health at the seaside and the extensive literature associated with therapeutic landscapes (Gesler, 1992; Andrews and Kearns, 2005; Foley, 2010). In health terms, some of this research draws on environmental psychology to identify aspects of reduced stress, attention-restoration and enhanced mood in parks, woods and forests (Hartig et al., 2015), while work by White et al. (2013), identified a measurable population-level health benefit from blue space, based on proximity to the coast and an associated maritime attachment. Gesler's (1992) identification of what he termed therapeutic landscapes, was initially built on empirical studies of healthy watering places such as spas, springs and wells and this has also been a feature of the author's earlier work in Ireland (Foley, 2010).

A more recent focus on blue-space geographies draws from both therapeutic landscapes and 'healthy nature' work to identify a range of water-based settings and how specific groups use them for healthy purposes (Foley and Kistemann, 2015). In Throsby's (2013) account of her experience of being a long-distance Channel swimmer, there were a series of thread's running through her account that were driven by specific challenges and important embodied relationships between training, diet and support. Representative coastal and inland landscapes and settings have been identified as providing benefits for physical and mental health.

Coleman and Kearns (2015), working with older people on the island of Waiheke (New Zealand) specifically identified an affective power and orientation to the water around this island. For some recently arrived residents, it was a mix of a surprise and long-held dream, but for people who had lived there a long time, a deep immersive experience on the islands beaches and bays across the years was central to their well-being. Even as their reduced bodily capacities made swimming or even getting to the beach more difficult, the view of the water and memories of those encounters, promoted a healthy resilience.

Tim Ingold noted that, 'through walking ... landscapes are woven into life, and lives are woven into the landscape' (Ingold, 2011, 47). In recolouring such a green quote blue, swimming lives have identifiable therapeutic accretive effects that thread together self and landscape across the lifecourse (Foley, 2017). Given this is an active, embodied and often emotional experience, the parallel effect is to produce what is often a very gentle and loving practice, within which an ethic of care, of self, of place, of water and of others is always present. Even if such practices of care are rare or infrequent, the thread, built on the relations and relationships that lie at the heart of relational geographies, is central to those connections. In mapping out that assemblage through case materials from Ireland and beyond, it cannot be denied that contested risk narratives are part of the story, but this chapter ultimately focuses on positive narratives of health and wellness and how these emerge in unexpected times, ways and places.

Swimming traces: approaches and settings

The (auto-) ethnographic material in this chapter is drawn from an ongoing set of projects from the last five years. These include an oral history of outdoor swimming in Ireland, drawn from three main sites. These were the 40 Foot on the south of Dublin City, the Guillemene in Tramore, County Waterford and the aforementioned Pollock Holes in Kilkee, County Clare. The interviews were relational in nature, bringing in narratives from other places and experiences from swimmer's lives. Most of the 20 interviewees (evenly spread across the sites) were over 60 years of age, an important consideration for the identification of life-course based blue threads. In addition, I have kept a personal swimming diary for the past five years, which also contains a glimpse of the past. Linked to this, I have been working on and with a number of visual technologies to try to work towards a closer understanding of *in situ* place responses. These include some spatial video recordings, using a technology developed at Maynooth University called *Ubipix*, which directly aligns video frames with GPS and digital maps and allows cross-format text tagging. It has the additional benefit of demonstrating the observer participation aspects of the study as well. Finally a range of secondary contemporary and historic sources have been used including, newspapers, old film, contemporary media and old photographs, to document that the places too have hosted decades of cross-generational practice.

With regard to settings, most are outdoor and by definition, exclude indoor locations, such as hotel swimming pools or public leisure centres. This is not to decry such settings and recent research is beginning to explore multiple settings and practices including indoor lap swimming (Ward, 2017). Though less memorable within mine and other narratives, indoor spaces do appear and remain part of an embodied echo. One final aspect of the empirical source material is that they are primarily positive. Yet there are some examples of bad experiences, which are important components of the story as well. Ranging from near drowning to theft and intense sunburn, these are all seared in the memory, though to date, I have not suffered the agony of a jellyfish sting, a constant fear for many mid-latitude swimmers (Rew, 2009). In the earlier mention of shared spaces I have equally been fortunate in my life to have learnt how to swim and had access to the water, even if it has been cold, and this is not something given to all (Collins and Kearns, 2007; Wiltse, 2007). On the whole, these experiences have been random and conditional on mood, time, freedom, company and inclination and I have nothing but admiration for those who are more disciplined and motivated to carry out everyday exercise whatever the season or swell. In the following section, framed by a generic life-course chronology, both personal and collective accounts are used to provide empirical evidence for the power and value of outdoor swimming as a relational act of health and well-being.

Swimming across the life course

Childhood immersions

I was brought up in County Laois, in the very centre of the country and as about as far away from the sea as one could get in Ireland. My first attempt to actually swim was in the nearby River Nore, in a shallow section where kids from my home town, Abbeyleix, went to learn (Figure 5.1). I did little more than paddle. But even before that I had had a scarier brush with the water. As ours was a middle-class family who took summer holidays by the sea, when I was around 4 or 5 years old we went to Courtown, on the east coast, where I wandered off into the shallows and was overturned by a wave. I can still feel the experience now, over 50 years later, a sense of impending doom and of helplessness in the power of the wave. Fortunately, I was fished out and finally learnt to swim during another family holiday, at the calm waters of the Tinside Lido in Plymouth. Through my childhood and early teens, it was almost always the west coast resorts of Kilkee and Lahinch that figure most prominently, mostly because they were shared with my parents and four siblings. All seaside resorts in Ireland (and indeed elsewhere) have a sort of catchment area. While Laois people generally went to Tramore, we went to

Figure 5.1 River Nore Pool outside Abbeyleix, Ireland.

Clare as my parents had worked there and had an ongoing affinity with the county.

The history of swimming at Kilkee is a lengthy one, and goes back in recorded archives to the eighteenth century (Foley, 2010). As the preferred resort town for the citizens of Limerick, who lived around 50 miles away up an estuary, it had a specific cultural history as a networked space. We were from a different culture but for that month of August, the shared space of the resort brought us into contact with that society and over time, that inter-mittent socialisation took on a magic of its own, as new friendships took root. While the sheltered beach at Kilkee was perfect for family swimming, the real focus was the Pollock Holes, a group of tide-filled pools in a rock–cut platform on the ocean's edge. There were three main pools, often referred to as the first, second and third, which respectively hosted family/children, female and male swimmers (Figure 5.2).

One of our first acts on the evening we arrived in Kilkee was to go for a swim, a bonding ritual for the family and a marker of our rush to get in the water. It always felt like the best swim each year and indeed, for the rest of the holiday we rarely all swam together in the same way. However, that swim was repeated by others across those summers to produce a layering of affective experience, swim by swim, day by day, summer by summer, family by family, which in turn provided a deep well of both practical and emotional identifi-cation with place and practice in one's later life. During my teenage years we tended to go to Lahinch (it had a better golf course) and by then I was also in

Figure 5.2 Pollock Holes, Kilkee.

a boarding school north of Dublin. At school, we were unusual in having an indoor swimming pool, where there was an emphasis on competitive swimming and life-saving. I failed my preliminary tests as the boy I was carrying was far heavier than me and dragged me underwater, and for that brief instant, the underwater flashback from childhood re-emerged. But the summer swimming was also hampered at Lahinch as it was much more open to the ocean than the protected waters at Kilkee. I preferred, at that time, to avoid the beach at Lahinch when there was any sort of swell.

Across many of the swimmer's stories, the role of either parents or a local informal coach who got kids started was noted by informants, as was a lingering familial bond:

> In Kilkee in 1947 I was 7 years old and I was taught by a nice man called Michael Duggan who taught all the local lads how to swim … swimming was in my family and I quickly became a proficient swimmer … at the age of 10, I swam across the big bay which was one nautical mile … one of the people who came with me was Richard Harris (the film actor) … he was about 19 at the time.
>
> (Respondent E, Kilkee)

> I think it's probably part of me … a part of me and equally my family, my sister does exactly the same … all three of us swim, even though one of them isn't that near to the water … we still swim a lot … and we love … I think we just …it's very difficult … I suppose it's something so

basic and its ingrained and it's there from nearly birth ... and becomes part of what you are as well.

(Respondent Y, 40 Foot).

The notion of swimming being 'in my family' was an important one as was the memory of who one swam with and to an extent, the sense of adventure and bravado that was often associated with the sometimes perilous practice of sea-swimming. I remember doing a Kilkee bay swim as well (the shorter 'little bay' crossing of about 500 metres) and feeling at the time like it was the most adventurous and dangerous thing I had ever done. Yet it was its very communal and relational nature, a life-event shared with 50 or so other kids, that helped me take that risk.

The soft dry years

For a lot of people, the arrival of adulthood, working responsibilities and other life events, often means that swimming, especially outdoors, can slip into the background. This is not always the case; the arrival of a child, for example, often brings families back to swimming pools and beaches. However, it was true in my case, wherein the only way of extending my swimming geographies was through typical trips during student holidays. One such trip brought me to the Greek island of Sifnos, where during a three-week stay with my sisters in a tent in the sand dunes, there was literally nothing to but swim. This was also the first prolonged exposure to wider dimensions of blue space, water and sky and crucially, warmth. Later as a young adult, I moved from Ireland to the UK, where I ultimately ended up in Brighton, one of the great seaside resorts of England. I never really took to swimming in Brighton as the rocky beach and relatively murky cold water didn't draw me in, though one still always gravitated towards the seafront. Even if my immersion in the seas off Brighton was limited, the draw of the water and its function as both a site and sight of well-being was always present and a constant of that time. But equally, there was a sense that the lack of specific immersive engagement felt like a softness on my part, a sort of shrinking away from my own body as it got wobblier and weaker, and as my own courage waned. For all that carrying out physical activity puts us in touch with our bodies, sometimes it confronts us with them as well (Throsby, 2013).

Other accounts talk about how personal and familial mobilities across time have the potential to both take people away from and bring them closer to water. Sometimes that movement is marked by life events. One Guillemene swimmer marked out a series of important events in his childhood and adult life that were marked by a swim; the loss of his father who had taught him to swim; a group swim in a lake on the morning of his wedding; and his bringing his own son to learn to swim on the coast north of Dublin. One 40

Foot swimmer had grown up swimming in a suburban river and had gradually made his way to swim along the coast to the 40 Foot, once he was old enough to swim in what was then a men-only swimming spot. As a young man, he became a professional footballer in Liverpool, but always came back in the summer to keep up that connection of swimming with his childhood mates. The idea of swimming spots as intergenerational spaces with multiple occupations across all ages is also captured by one informant:

> Yeah it would be very mixed actually ... really you'd have everyone from kids up to people in their 80s ... I suppose maybe it's maybe the older crowd who tend to spend a lot of time down there and the younger crowd tend to come and go a bit more.
>
> (Respondent N, 40 Foot)

A rediscovery of water

My return to Ireland took me to Dun Laoghaire, on the southern end of Dublin Bay, where the 40 Foot swimming spot acted as a magnet, initially for observing the sea and swimming, but gradually drawing me in to occasional practice. My new partner Nell was part of a long-standing 40 Foot swimming family and gradually I got to know a range of swimmers, both regular and intermittent (of whom I was one), at that place. A few years before, my parents had retired to Tramore and this brought me in touch with the swimming community at the Guillemene, where I became a country member of the Newtown and Guillemene Swimming Club. In both swimming at and interviewing swimmers in these and other places, a range of often-poetic responses to the water emerged. One respondent compared entering the water at the Guillemene to being like the shift from black and white to colour in the film version of the Wizard of Oz, while also commenting on the colour of the water: 'It depends you see on the day ... the water could be brown, it could be aquamarine, it could be green, it could be vivid blue, it could be ... if the sea is agitated it could have sparkle' (Respondent E, 40 Guillemene).

As I recorded my own intermittent practice, the relationships between time, season and practice emerged in the words of swimmers answering a question on how long they swam for:

> Yes obviously .. in the summer it can go up to, I can stay in the water for up to an hour ... but right now – this is still June ... you could stay in the water for about half an hour ... in winter I'd be lucky if had 60 seconds in the water. The exercise is in getting dressed and undressed I think (laughs) more than the swim ... (Q. *So you don't make a distinction*

between the different experiences, quick dip and long swim?) . . . no because I think they change with the seasons, it's a gradual thing and . . . you can have a swim as opposed to a dip up till Christmas and then usually January and February are the coldest months for swimmers . . . I mean the water is quite cold until May . . . so the middle times it's teabag swimming.

(Respondent Y, 40 Foot)

Another aspect of outdoor swimming in colder latitudes is a certain embodied machismo about whether one wears normal togs or any sort of wetsuit. In my childhood, one always swam in togs, but looking at beaches now, it is rare not to see a child in a shortie wetsuit. At the 40 Foot and Guillemene, wearing a wet suit, especially among the regulars is a definite no-no and as one of the former noted, 'there has to be a chill to make you feel good'. I compromise by swimming with a thick 3 mm rash vest, a sop to my combined cowardice and weak immune system. As another colleague argues, if wearing a wet suit in cold water enhances your ability to swim farther and stay in longer, surely this can only be a good thing.

Clearly, the experience of warm water swimming, especially on a sunny summer's day, remains an appealing contrast. The extract below, of a swim off the Greek island of Hydra, was chosen in part because of the clarity, depth and warmth of the encounter, an echo of earlier and later plunges, but also that sense of the liminality of the experience in that uncertain zone between land and sea, even where that interface wasn't tidal.

Coming in to the enclosed harbour, I could see gangs of people swimming to the right and went around there as quickly as I could. There was something about the location, the view, the cleanliness of the water and the contrast between the white houses and blue depth of colour in the water that would have made anyone want to dive in; indeed, there was almost no other way to get in. In fact, the water had quite a swell, amplified by the swash of boats coming in and out to the harbour. Diving in, one was immediately drawn around to some caves under the limestone rocks that marked a sort drawing in under the land, a permeable itch that the body almost naturally followed. However, the water was very deep, very clear and one would have happily stayed in there for hours.

(Swimming Diaries Extract: Hydra, Greece, 30 August 2011)

More recent auto-ethnographic work has focused on a more direct measurement of the experience of swimming using a variety of new technologies that seek to explore the *in situ* nature of human–water interactions. Figure 5.3 shows a still from a recent Ubipix recording of a swim at the 40 Foot, using a spatial video tool that recorded the walk to the swimming spot,

Figure 5.3 Still from Ubipix recording of 40 Foot, swim by author, 30 October, 2015.

went up to the edge of the water and then recorded the swimmer (myself) in the water and on the way out, where a series of squeaks, grunts and gasps recorded a rather inarticulate though definitely embodied response to the whole experience, given the coldness of the water in late Autumn. What the Ubipix software does, however, is also convert the movement in the video into a set of GPS-tagged tracks so one can see the trace of the walk into the blue alongside the orientation of the camera and measures of the height and distance covered in the video. This spatial video technology also allows one to mark or tag specific points from the image and give them additional tags, either as direct *in situ* transcript texts or subsequent elicitation commentaries, but the technology is still at an experimental phase.

Finally, Figure 5.4 shows a second example of experimental *in situ* work. The image records a swim-along interview test at Killiney Beach, where the interviewee was recounting the history of an old beachside teashop, when a long-distance swimmer suddenly appeared and silently passed. These intermittent interruptions often form part of a swim experience and are part of the always uncertain outcome of any entry into the water. It is as if, by giving over control to the water, we lose a certain amount of

Figure 5.4 Still from swim-along interview recorded on action camera, Killiney, July 2015.

control over what might happen in there as well, even if we do not feel necessarily unsafe.

Discussion: swimming as relational healthy practice

In trying to consider swimming as one way of understanding embodied and relational human–water relationships, one can see a range of fluid palettes and practices at play. Many languages, Vietnamese and Irish among them, have a fluid understanding of what the colours green and blue are. In Vietnamese, the use of the word *xanh* is usually followed by the question; do you mean *xanh* like the trees or *xanh* like the sea? (Regan, n.d.). In Irish the word *glas* has similarly ambivalent qualities with hints of added greyness as well; especially appropriate in a climate where the colour of the sea changes 20 times a day, though we continue to try and represent this intermittent colouring as blue (Loeffler, 2015). Artists too have been drawn increasingly to and capture this experience with the Irish swim art of Vanessa Daws and her video-based Lambay Island triptych an example of the ways in which active practice and creative expression combine together, something she characterises as 'psychoswimography' (Daws, 2017). Here the role of swimming places as affective relational settings for health and well-being are always evident.

In uncovering some personal life-course narratives, there are stories of health risk – unhealthy and dangerous experiences and practices – such as people jumping in when it is too shallow or rough, an example being the photography of Martin Parr and Seamus Murphy (Parr, 2009; Murphy, 2017). But more importantly, most people describe the fundamental joy and value of swimming. Many of the voices in this chapter look back on old baths and beaches that carry with them echoes of a past that is gone, but lives on in affective memories buried deep in bodies. The ways in which different swimmers occupy and own swimming spaces is also reflected in that relational and shared history. Wider research has identified swimming pools and beaches as sites of exclusion for people of colour, while other aspects of belonging and the ways in which a freedom to swim operates in different ways and settings around gender and class (Wiltse, 2007; Bolton and Martin, 2013; Lobo, 2014). This is a particular focus for the new 'wild swimming' movement and intriguingly, recent evidence suggests that the practice of outdoor swimming is growing at the expense of indoor pool swimming, the first time this has occurred in decades (Laville, 2015).

But equally, whether indoors or out, there is something about the act of swimming and the repetition involved that seems to express itself as a body knowledge that never quite goes away and becomes, as a pre-cognitive body act, an affective capacity. Much of the recent academic writing around water and health focuses on the three 'Es', experience, embodiment and emotion (Bell et al., 2015; Foley, 2017). In those accounts and the ones included in this chapter, aspects of flow and depth emerge as recurrent terms linked to the uncertain displacement of embedded affects, mostly pleasurable, but sometimes with fearful echoes. In writing a swimming diary and interviewing older swimmers, there are common references to going back to being a child again. My father passed away in 2013, a slow fading of his embodied presence but in thinking of him as a swimmer I remember him differently. He had a slow steady powerful swimming stroke, which was slightly at odds with his light place in the world. This glimpse into a guarded life echoes my own as another reason to write a swimming diary and to swim in his wake. Swimming can be a way of letting down your guard, letting the outside in an embodied way, but equally keeping it out mentally. My father's own love of the sea was passed on by his mother who, unusually for a young woman in Ireland in the 1930s, had a car and took him on trips to Ballybunion and Ardmore. This act of passing on a love of water and swimming is part of an almost genetic blue thread that runs across relations and generations that in turn is translated into a wider ethic of care around water that transcends familial bonds; as witnessed below in a vignette observed at the 40 Foot:

> Then saw a quite old man with walking stick hobbling down to steps and observed a nice episode of care as he made his way along clutching railings and manoeuvring himself past people already on the steps. Then a woman – possibly Ciara – offered to look after his stick while he was in the water. He accepted the offer but said he wouldn't be in long. He

manoeuvred his way past a man who knew him – 'how are ya Frank' and then went in and swam around 60–70 metres with a heavyish slappy overarm stroke and then came out past a father and son going in and was met on the steps with the stick and then back up to land … almost seemed like there was a little extra spring in his speed of step as he went back to his clothes.

(Swimming Diary Extract: 40 Foot, 7 June 2015)

Conclusion: replenishment and restoration

A recent video on the BBC website captured a group of older swimmers in Porthcurno (Cornwall) (Guardian Online, 2017). The swimmers spoke of swimming as a form of erasure and invoked the sense of crossing a threshold and a deliberate unfolding away from workaholic lifestyles. They also identified a sense of being 'out of shape' before the swim but that after even a single swim, that act alone restored a healthy balance and a sense of renewal, echoing environmental psychology writing on restoration (Hartig et al., 2015). While many of the voices in this chapter belong to regular swimmers, I would argue that it doesn't necessarily matter how far apart the swim is; if anything the affective response and the tug on the thread feels more sharply articulated after long gaps. This in turn invokes a relational geography operating across the life course that is not so much a single thread as an irregular web that replenishes and restores itself. That geography articulates the meeting points as the swimming spots and actual swims, the lines of connection as the more embodied and socially produced supports. While the process can see a thinning and thickening of the filament, it somehow holds together, echoing the astonishing strength of a natural web that is structurally much more powerful than it appears. As an intermittent yet accretive process, swimming is an exemplary example of the creative ways in which humans and water interact for the specific promotion of health and well-being,

References

Andrews, G. and Kearns, R. (2005) Everyday Health Histories and the Making of Place: The Case of an English Coastal Town. *Social Science & Medicine*, 60, 2697–2713.

Bell, S., Phoenix, C., Lovell, R. and Wheeler, B. (2015) Seeking Everyday Wellbeing: The Coast as a Therapeutic Landscape. *Social Science & Medicine*, 142, 56–67.

Bolton, N. and Martin, S. (2013) The Policy and Politics of Free Swimming. *International Journal of Sport Policy and Politics*, 5(3), 445–463.

Coleman, T. and Kearns, R. (2015) The Role of Bluespaces in Experiencing Place, Aging and Wellbeing: Insights from Waiheke Island, New Zealand. *Health & Place*, 35, 206–217.

Collins, D. and Kearns, R. (2007) Ambiguous Landscapes: Sun, Risk and Recreation on New Zealand Beaches. In Williams, A. (ed.), *Therapeutic Landscapes*. Farnham: Ashgate, pp, 15–32.

Daws, V. (2017) *Lambay Circumavigational Swim*. http://vanessadaws.com/lambay-cir cumnavigational-swim/.

Deakin, R. (1999) *Waterlog. A Swimmers' Journey through Britain*. London: Chatto & Windus.

Foley, R. (2010) *Healing Waters: Therapeutic Landscapes in Historic and Contemporary Ireland*. Farnham: Ashgate.

Foley, R. (2017) Swimming as an Accretive Practice in Healthy Blue Space. *Emotion, Space and Society*, 22, 43–51.

Foley, R. and Kistemann, T. (2015) Blue Space Geographies: Enabling Health in Place. Introduction to Special Issue on Healthy Blue Space. *Health & Place*, 35, 157–165.

Gesler, W. (1992) Therapeutic Landscapes: Medical Issues in Light of the New Cultural Geography, *Social Science & Medicine*, 34, 735–746.

Guardian Online. (2017, 13 February) The New Retirement. *Guardian*. www.theguar dian.com/society/video/2017/feb/13/wild-sea-swimming-in-my-60s-erases-pro blems-being-child-again-video.

Hartig, T., Mitchell, R., De Vries, S. and Frumkin, H. (2015) Nature and Health. *Annual Review of Public Health*, 35, 207–228.

Ingold, T. (2011) *Being Alive. Essays on Movement, Knowledge and Description*. Abingdon: Routledge.

Jones, O. (2011) Lunar-Solar Rhythmpatterns: Towards the Material Cultures of Tides. *Environment and Planning A*, 43(10), 2285–2303.

Laville, S. (2015, 18 August) Different Strokes: Open-Water Swimming Takes the UK by Storm. *Guardian*, S2, p. 4.

Lobo, M. (2014) Affective Energies: Sensory Bodies on the Beach in Darwin, Australia. *Emotion, Space and Society*, 12, 101–109.

Loeffler, S. (2015) *Glas Journal*: Deep Mappings of a Harbour or the Charting of Fragments, Traces and Possibilities. *Humanities*, 4, 457–475.

Lorimer, H. (2015, 20 August) Atypical Situation. *Environment and Planning D: Society and Space*, Online Essay. http://societyandspace.org/2015/08/20/atypical-situation-hayden-lorimer/

MacEvilly, B. and O'Reilly, M. (2016) *At Swim: A Book About the Sea*. Cork: Collins Press.

Macfarlane, R. (2010) *The Old Ways. Journeys on Foot*. London: Penguin.

Mitchell, R. (2013) Is Physical Activity in Natural Environments Better for Mental Health than Physical Activity in Other Environments? *Social Science and Medicine*, 91, 130–134.

Murphy, S. (2017) *The Swimmers*. www.seamusmurphy.com/Photography/The-Swim mers/1.

Parr, M. (2009) *The Last Resort*. Stockport: Dewi Lewis.

Parr, S. (2011) *The Story of Swimming*. Stockport: Dewi Lewis.

Regan, N. (n.d.) *Grue*. unpublished poem.

Rew, K. (2009) *Wild Swim*. London: Random House.

Robinson, T. (2012) *Connemara: A Little Gaelic Kingdom*. Dublin: Penguin Ireland.

Seamon, D. (2013) Lived Bodies, Place, and Phenomenology: Implications for Human Rights and Environmental Justice. *Journal of Human Rights and the Environment*, 4(2), 143–166.

Sherr, L. (2012) *Swim. Why We Love the Water*. New York: Public Affairs.

Throsby, K. (2013) 'If I Go in Like a Cranky Sea-Lion, I Come Out Like a Smiling Dolphin': Marathon Swimming and the Unexpected Pleasures of Being a Body in Water. *Feminist Review*, 103, 5–22.

Ward, L. (2017) Swimming in a Contained Space: Understanding the Experience of Indoor Lap Swimmers. *Health & Place*. online corrected proof.

Ward-Thompson, C. (2011) Linking Landscape and Health: The Recurring Theme. *Landscape and Urban Planning*, 99, 187–195.

White, M., Alcock, I., Wheeler, B. and Depledge, M. (2013) Coastal Proximity, Health and Well-Being: Results from a Longitudinal Panel Survey. *Health & Place*, 23, 97–103.

Wiltse, J. (2007). *Contested Waters. A Social History of Swimming Pools in America*. Chapel Hill: University of North Carolina Press.

Wylie, J. (2005) A Single Day's Walking: Narrating Self and Landscape on the South West Coast Path. *TIBG*, 30(2), 234–247.

Chapter 6

Creative compulsions

Performing surfing as art

Jon Anderson and Lyndsey Stoodley

Introduction

This chapter focuses on how one particular water scape creates an attractive force that 'ensnares' and comes to define human beings: the surf zone. The surf zone is a geographical space produced through the meeting of land, sea and air. It is both a process and a product of the coming together of these three physical media, and, if this alchemical admixture comes together in just the 'right' way, then a liquid gold – commonly known as surf – is created. These cresting and breaking waves of water offer the opportunity to glide, slide or otherwise ride the energy that passes through the surf zone. By drawing on non-representational ethnographic work (after Vannini 2015) from surfers from around the globe, alongside auto-ethnographic experiences (after Ellis 2004) of the authors, this chapter builds on existing work on surfing identities (Evers 2009; Usher 2015; Olive 2016) to explore why many people rush to engage with this liquid gold, choosing to engage with it directly through surfing on boards, boats or simply with the body. This chapter suggests there are three, often interrelated and interchangeable, ways in which individuals articulate their identification with the coast, beach and surf zone: first, individuals may define themselves in relation to a specific beach and its coastal waters in a 'surf-shore identity'; second, they may articulate an affiliation with the abstract liquid territory beyond terra in a 'water-person' or 'sea-space identity'; and third, they may define themselves through specific engagement with the moment of surf-riding, an identity created through 'emerging performance'. It will be argued that this latter fluid category underpins many surf-riders' identities and produces what Chamberlain (1996) terms a magnetic attraction to surf; this chapter develops this assertion through adopting the language of the medical humanities and introducing the idea of a 'compulsion' to engage in surfing. The chapter will suggest that the notion of compulsion helps us to art-iculate new ways in which we can understand the practice and performance of the surfing act.

Surfing selves

As has been noted elsewhere (Sack 1997; Casey 2001; Anderson 2015a), humans are spatial beings, their identity and sense of self is co-created through engagement with the physical, social and cultural world. This emergent co-creation occurs through daily experience and interaction (see also Anderson 2004, 2014a); as the human body physically encounters places (through a process Casey calls 'outgoing'), traces of location are inscribed on the human self by 'incoming' strata of meaning (see Casey 2001, 688). As this process occurs, our own practices reciprocally come to influence the meaning of the physical, social and cultural locations we inhabit (see also Halbwachs 1992; Crang and Travlou 2001). Despite geographical and social science traditions often valorising such co-creative engagements in the terrestrial world, the identities of many humans are also formed by their connections with territory beyond terra (see Peters, Stratford and Steinberg 2017). Water worlds play a significant role in shaping our selves, and vice versa (see Anderson and Peters 2014; Brown and Humberstone 2015). Indeed, a growing body of social science research recognises and explores how humans begin to co-emerge not only with the water world in general, but the coast, beach and surf zone in particular.

Human co-emergence with this specific aspect of the water world has been understood in a number of ways. As Anderson (2014a) argues, humans can be defined as much by their relations and practices to particular coastal shores as well as the surf zones lying beyond them. This 'surf-shore' identity is most commonly created at a specific spot where humans engage with surfing waves on a regular basis. The 'local' beach with its own idiosyncratic customs, cultures and power relations becomes known to the individual, as does the more-than-human processes and agencies of the surf break itself. Individuals develop an awareness of where waves break in particular conditions, how the waves change as weather systems and tides pass through, and these hard-earned insights accumulate with personal and collective experiences over the days, months and years engaging with this singular littoral geography. This regular practice 'sutures' (after Heath 1981) the individual to the beach and surf zone of their local; this 'surf-shore' identity position is described first by Evers, then Brooks:

> Surfers form relationships with the local weather patterns, sea floors, jetties and rock walls. The coastline and weather have enormous personal meaning for us. There is a close link between physical geography, stories and bodies in the act of surfing. The dynamic, unpredictable energy of the waves and coastlines and weather patterns flows through our bodies.
>
> (Evers 2010, 47)

> You can surf all over the world, you can score dream sessions at even dreamier waves, but nowhere will shape you quite like your local. It's the

place you come to know better than any other, that you love and hate and learn from constantly, and it leaves an imprint on the salty side of your soul that can't be erased.

(Brooks 2017)

These quotations suggest that regular engagement with a particular surf beach and break can influence how individuals sense their world, and by extension their own relations to it. From this regular engagement a hybrid identity is created that goes beyond the land and connects lives and practices to the littoral zone. In other words, a 'surf-shore' sense of self is accreted that comes to define an individual who dwells in this location. However, through engaging with the sea, perhaps from a particular beach or coastal location, or simply from any surf-side site, some individuals express their identity in subtly but nevertheless significantly different ways. For these individuals, the lure of sea-space itself is fundamental; they feel most 'at home' in the littoral zones beyond the shore. Such individuals express themselves as 'water people' (see Mattos 2004), or people of the 'sea-space' – it is in the sea that they are in their element. Corporeal engagement with the energy of waves has such a transformative, often therapeutic, effect on their being that this becomes the space to which they feel most affiliated. As the following surfers articulate:

Surfing clears my mind and forces me to focus on what is going on around me: other surfers, approaching waves, currents and rocks, being in the right place to catch waves. When all this works together to catch a wave the feeling is indescribable, and this is the reason I continue to surf.

(Survey Respondent 2015)

Surfing is [now] a huge part of my life, every surf session feels like I'm being cleansed, I love nature and the natural beauty of the waves and the joy they bring to us all!

(Survey Respondent 2015)

It is a place of healing, of peacefulness, self-reflection, and inspiration. I also teach younger students how to surf and the pure joy and confidence that surfing brings into their lives is remarkable. Spending time in the water brings happiness, it feels safe to me and being on a board connecting with nature makes me feel elated.

(Survey Respondent 2017)

I was born to surf. It's what my soul needs to stay balanced while on earth. [It allows me to overcome] physical, mental and emotional challenges. I use surfing as my platform for problem solving, it's the best tool. There's no aspect of the surfing process that I can't use to relate to everyday life issues.

(Survey Respondent 2017)

These articulations suggest that surfing enables some individuals to define themselves in relation to surfing 'sea-space'. Indeed, engagement with this element enables them to escape the terrestrial versions of their selves (perhaps even their specifically located hybrid surf-shore identity if they have one), and express a new iteration of their self in relation to the generalised littoral zone (see Anderson 2014b). Indeed, these articulations of 'surf-shore' and 'sea-space' identity resonate with social science understandings of how poststructural identities exist and proliferate. As Wood (2012, 195) has noted, from the mid-1980s the fixed idea of a unified identity has given way to acknowledgement of 'dynamic, multiple and fractured identities'. With this in mind, surfers may not simply have one surf-shore or sea-space identity, but rather express a multiplicity of selves (Jameson 1991; Featherstone 1995), with each articulated in a different geographical and cultural location. It could be suggested that surfers therefore have a spatial – or in this case perhaps even an elemental – *division* of identity (see Anderson 2004, 2014a), with a portfolio of not necessarily coherent selves being expressed in different media and sites. What is crucial to these individuals, however, is that engaging with the surf zone is an essential aspect of this portfolio, an experience that brings balance and homeostasis to an overall whole, which would be discordant and unmanageable without it. As Evers (2010: 54) puts it, '[If] I haven't surfed for a couple of weeks [, work gets] the better of me. My body lets me know when the break between sessions has been too long. I start to feel restless and get annoyed over little things. My temper becomes short'. For these individuals they can only be who they are in aggregate *because* of their singular engagements with water.

We have seen thus far how processes of 'incoming' and 'outgoing' suture individuals to the sea; their relations with water come to define how many surfers sense their selves. It is important to note as this stage that these identity positions are not fixed and mutually exclusive states; it is possible, for example, for a surfer to articulate both a 'surf-shore' and a 'sea-space' identity at different times and in different places. Coupled to this, these identity positions are not stable or absolute in nature. Reflecting their connection to the fluid media of the sea, these multiplicities of identity do not arrive fixed and formed in a full and finished state, but rather are produced through processes of co-creation with people and locations (for more, see Williams 1977; Wood 2012). This acknowledgement of the *emergence* of identity (see Haraway 2003; Anderson 2009) places emphasis on the moments in which identities are called forth, and their consequences; it draws attention to the instances of interaction and their aftermath through which new versions of ourselves emerge through practice (see Butler 1990, 1993). This focus on the emerging present, on the momentary now and its consequences, is central to turns to the performative within the social sciences (see, for example, Nash 2000; Turnbull 2002; Holt 2008). As suggested by Sutton-Smith (1997) and Pratt (2000), these turns emphasise how identities *emerge* through, and only through, spatial practice. It is only in the moments of play and performance – in other words in those moments when practice is undertaken – that identities are temporarily realised.

If we apply insights from the performative turn to the practice of surfing we see the act of surf riding itself, and its consequences for human relations with the surf zone, differently. In relation to the subject(s) that concern us here, surfing identities only come into being within the precise moments of their littoral engagement. Such perspectives suggest, therefore, that a surfer is only a surfer in the moment of surfing. One may be a water person through 'simple' engagement with the sea, one may articulate a surf-shore self by identifying through practice with a particular beach location; yet to become a surfer suggests a particular engagement that has to be performed to be enjoyed. Regardless of previous attainment or personal experience of having ridden waves successfully in the past, each cresting wall of water offers an opportunity for the surfer-self to be momentarily realised, and with each wave that breaks unridden the potential of becoming a surfed wave and a surfer is lost to the water and the individual (unrealised) rider (Figure 6.1).

From this perspective, therefore, an individual is defined by the act they are engaged in at that moment. In relation to surfing, every encounter with

Figure 6.1 The becoming, and unbecoming of a surfer (Stoodley 2017).

Figure 6.1 (continued)

the surf zone becomes a test or ritual for the human who wants to realise a momentary surfer identity. If one does not position one's self correctly, one (or one's many potential selves) do/es not emerge as a surfer. If one wipes out (or crashes from a breaking wave), one does not emerge with a surfing experience. The surfer self can only be called forth in the moment, and momentarily. If we take this performative interpretation of identity seriously, we come to a situation where even though in common parlance we may refer to individuals as 'surfers', we are more precisely suturing an identity label to individuals that is historical in nature – they surf, or more accurately they have surfed in the past. In the current moment, unless they are actually engaged in the act of surfing, they are not surfers. This apparently pedantic exercise in linguistic accuracy is actually more important than it appears. First, it is significant as it draws attention to the elusiveness and rarity of successfully achieving a surfing identity. One has to wait for appropriately assembled land, sea and air conditions, be well positioned at the correct place on the water surface, and bring together combined experience and skill to ride the probably unpredictable and eminently uncontrollable energy surging through the molecular structures of the sea. When the state of surfing is attained, its affective power is felt strongly by the successful rider – the thrill, rush, fear, satisfaction and joy (often all rolled together in the word 'stoke') is the product of the unlikely convergence of sea, swell, solar energy, continental shelf, beach, wave, board and rider. It is this relational sensibility, this rush of stoke, that rewards the patience and dedication of the aspiring surfer. Second, and relatedly, drawing attention to the performative moment of surfing also helps us acknowledge the absence an individual feels when they are *not* engaged with the surf zone. Although the after-affects of stoke can be remembered and re-presented through story-telling and memory, these processes do little to replicate the sensation of experiencing the successful convergence with the breaking wave, and achieving the emergence of a surfer-body. Indeed, this re-packaging of the surfing experience through stories and images (or even through broader commodification by surf companies and media, see Reilly 2003) often simply reminds individuals that there is nothing quite like the emerging moment of surf-riding. These reminiscences and re-presentations often have the affect of charging their wish for re-engagement with the sea. Emphasising the performative therefore reminds us that many indivi-duals who have an aspect of their identity co-constituted by and with water are not who they wish to be when they are away from seaspace; they experience a sense of exile from the element that has come to define them. In short, the momentary versions of themselves that emerge when they are in the (surf) zone are revealed to be so enticing and powerful that these individuals do not feel 'right' when they are not directly experien-cing them. As Zink (2015, 81) articulates:

> Moments when things feel right draw me back to the sea. The sea is a place where that feeling occurs in a way it does not anywhere else. Maybe what draws me back is that the assemblages that form at sea work to loosen the borders of identity and expand both perception and the capacity to enter into relationships with other bodies in new and interesting ways. I go to sea because it affords opportunities to be in relationships that 'feel right'.

Creative compulsions

If we turn to nascent studies in the medical humanities, we can understand these moments that 'feel right' in a new context. Studies with individuals diagnosed with conditions such as Tourettes Syndrome or Obsessive Compulsive Disorder (see Ricketts et al. 2012) demonstrate that individuals only ever define their relational sensibility as 'just right' when they act on a chemical, psychological or bodily insistence to engage in a particular behaviour, such as rearranging objects in a particular order or interacting with items in repetitive ways. When interpreted geographically (see Rasmussen and Eisen 1992; Beljaars forthcoming; Beljaars and Anderson, forthcoming), such individuals define their person–place relations as only ever 'just right' when they affirmatively surrender to their compulsive urge to rearrange the relations between their bodies and the material objects in place round them (for example, they re-order books on a shelf, touch an ornament in a certain way, or have their curtains pulled to a specific degree). In short, they only emerge with an identity that feels 'just right' when they act on their compulsion to engage with their world in a certain fashion. Due to the ways in which some water people articulate similar 'just right' tendencies with respect to their engagements with the sea, it is provocative to consider surfing in light of these framings. It could be suggested that surfers' relationship to their material world could be understood to have a 'family resemblance' (after Goffman 1977) to compulsive traits. For those with a defining affinity to water worlds, their person–place relations will only feel 'just right' when they affirmatively surrender to the magnetic pull of the sea, and the compulsion it creates in them to surf. It is only when they do so that they can feel 'just right': it is at these moments that they can experience the in-the-now emergence of their surfing identity. This state is implied by the remarks from Evers (2010, 54), but is also expressed in the following remarks made from individuals who are not currently emerging as surfers:

> [surfing is] a seriously addictive passion. Initially I had assumed that surfing once a week or fortnight would be more than sufficient. Yet the more I surfed, the more I wanted; the hungrier I became.
>
> (Capp 2003, 82)

> Once you've had a taste of [surfing, it is] something different, something of out there, [and] then it's hard to give it up. [It] gets its hooks in you. Afterwards nothing else can make you feel the same.
>
> (Kotler 2006, 153)

I'm sat here at my desk, the sun is out and, by habit, I checked the surf forecast then the webcam at my local beach (see Figure 6.2). This was

Figure 6.2 Concentrating in class, on the next potential surfing opportunity (photo L. Stoodley 2017, charts BBC weather 2017 and Magicseaweed 2017).

foolish; I am now totally regretting not getting my work done earlier and wishing I was there. I know exactly what it will be like, I can see it if I close my eyes, and I can feel the sun and water on my skin. On a sunny day like today the water glistens, the wind is low and the sound of the waves puncture the peaceful serenity of a weekday beach. People will be few and far between, until the workers start to trickle into the car park – 4 o clock for those who work locally. Those who, like me, work an hour away from the beach will start appearing a little later and rush to get into their wetsuits to make the most of the fading light. I look to see what tomorrow is likely to be like, to reassure myself. It's okay, you can go tomorrow. It's fine. Really.

Really though, it's not fine and now I want to drop *everything* and head west to those little waves that are breaking lonely onto a sunny beach.

(Stoodley 2017)

> Surfers . . . need their fix, if the government knew how addictive surfing is they would have outlawed it by now.
>
> Leza (2012)

With these insights from the medical humanities in mind, we can begin to explore the idea that for some individuals, surfing can be defined as compulsive act. In order to feel 'just right', aspiring surf riders have to affirmatively surrender to their compulsive urge to bodily engage with the water through the act of wave riding. This position provides a new lens through which to understand the practice of wave riding, and what it means for the management and articulation of identity. As outlined in the passage above, the magnetic draw of the surf strongly tempts individuals to change their working habits in order that they can engage with the littoral zone; and as the following excerpt from respected surf magazine *The Surfers Path* outlines, there are a range of other actions that aspiring surf riders engage with in order that their lives can be rearranged around their compulsive urge to surf:

> Well for a start, [those with surf compulsion] have probably made some major, life-changing decisions around the need to live near the ocean and to somehow create enough freedom to surf every day good waves are happening. [They] probably can't look at any piece of water, north or south coast, lake, river or stream, without searching every detail of the water surface, the bottom contour, wind and swell for the sign of rideable waves. Even if they're six inches not six feet, there's still something irresistible about them. [Their] partner probably has a problem with [their] obsession and passion for the sport: 'If only you were into me as much as your damn board!' [Those with this compulsion] can't contemplate never being able to surf again. The prospect of the next great day and [their] next great wave is one of the things in life that keeps [them] going forward. Moving away or just not being physically capable of surfing are prospects that just don't bear thinking about. [They]'re always restless, dream[ing] of the next session, [their] optimism tells [them] that the next swell will be the best ever, and [they]'ll be on top of [their] game when it comes. On the days it's working but [they] miss it, [they] get eaten away inside with frustration. [They]'ve probably bunked off school or work many times because it really was too good to miss. Driving to good surf, hurriedly changing and running headlong to the water, fills [them] with stomach tightening anticipation, the same as it ever did.
>
> (Richards 2002, 114)

In this view, surfing could be interpreted as a compulsive act – an articulation of affirmative surrender to the lure of the littoral. In order to experience the relational sensibility generated by successfully riding waves and thus become, momentarily, a surfer, individuals must engage in this activity directly. This 'just right' feeling is so magnetic, so ensnaring to the aspiring surfer, that they feel an overwhelming urge to prioritise this practice, and thus this aspect of their identity, over all others. In the same way that an aspiring writer must write, or a wannabe sculptor must sculpt, an aspiring surfer must surf.

Art-iculating surfing selves

The act of surfing thus may be considered an act of compulsion, and at once an articulation of surfing identity. Through its practical expression, a statement of self is made, and therefore the act of compulsion is also one of creativity: the surfer expresses who they are and their relations to the watery world around them through their presence and actions. Such framings lead many to acknowledge that surfing practice is not just a passive expression but more a creative articulation of identity. As the following surfers describe:

> Surfing is an art. The metaphor goes like this: the board is the brush, the wave is the canvas, and the surfer is, of course, the painter ... Like in art, each artist is unique. Each stroke of the brush is an expression of an individual style, mood, and life experience.
>
> (Tess 2011)

> The wave was a canvas and they painted it with whatever was in their imaginations and bodies to paint.
>
> (Heller 2010, 53)

Through expressing an integral aspect of their identity through surfing, the surfing moment becomes a way for individuals to articulate their thoughts, feelings, and outlook on their person–place relations, in and through practice. In this (perhaps romanticised) view, surfing is more than simply a sport or a hobby, but a means through which humans can artistically express and become themselves. In the same way art takes many forms, surfers take to a wave with differing styles and aims, in line with their chosen discipline. Those riding short boards, generally around 6 feet in length, may seek to ride a wave aggressively, 'shredding' the face of it with big, deep turns that spray water out of the wave. They may even attempt to leave the wave altogether and add an aerial element to their performance. By contrast, longboarders, whose craft are 9 feet or more in length, will generally move in a calmer, perhaps more languid, fashion, aiming to cross step their way to the nose of the board to dangle their toes in the wave, or 'hang ten'. Whatever form this expression takes, the human–water context in which these selves are articulated lend a particular nature to the creative practice. The selves that are articulated in this context are not discrete and isolated, they are constituted

and composed by the assembled components of board, swell, wind, geomor-
phology, sea (etc.) that contribute to the relations producing the surfer
themselves. Similarly, the 'art' that is produced is co-composed by this
assemblage, it is not simply the work of the 'painter' (or surfer) in isolation,
but authored too by the moving processes in and through which the surfer
emerges. The authorial (art)work of a 'surfed wave' is, in the terminology of
Ingold (1993), a 'taskscape'. It is the assembled process of skilled practices
articulated in an environment. Yet this is not a taskscape in the sense of the
production of a definitive and durable product that can be witnessed and sold,
but rather the creation of a process that is transient and ephemeral. As Ingold
(1993, 161) describes with respect to the task of painting, 'The emphasis,
here, is on painting as performance. Far from being the preparation of objects
for future contemplation, it is an act of contemplation in itself.' With respect to
surfing, the experience of carving a unique line through a once in a lifetime
coming together of wind, sea, rock and swell can be resonant with the idea of
creating an ephemeral taskscape. In this view, surfing is not simply a practice of
recreation, more accurately it is a process of (re-)creation of person and place as
art. Despite the burgeoning popularity of surf film-making rendering it possible
for this taskscape to be captured and crystallised by the surfer themselves or by

Figure 6.3 Carving the canvas (L. Stoodley 2012).

others witnessing its creation (as we can see in Figure 6.3), the real power of this taskscape can only be experienced by the surfer who is creating it. As Heller suggests, surfing is in this sense 'an aesthetic act[,] a performance art that requires no audience[, a composition of] painted arcs done without a brush upon an every-changing canvas' (Heller 2010, 53); and as the following surfer describes, the relational sensibility derived from performing and creating this artistic taskscape adds further addictive dimensions to the stoke potential of the practice:

> I feel at peace [when surfing]. It's the only time my mind shuts off and I'm just thinking about the next wave, the wave I'm on, and not drowning if I get caught in a bad spot (!). When I'm on the wave I feel *so* connected to nature, it's doing it's thing and I am interacting with it, reacting to what its doing, and where it's going.
>
> (Survey Respondent 2015)

As we have seen from the chapter thus far, this in-the-moment performance of the task, of a barely perceived act of creation that is both fleetingly lived and glimpsed, becomes part of the individual's identity, a source for reminiscence, and a motivation for continued attempts to engage successfully with the surf zone; as Richards notes:

'You carry the memories of [the] many magic days, shared with the best of friends. [The] images of falling lips, light offshores, late drops, speeding walls, barrels, sunsets and big gnarly days [become] etched on your brain' (Richards 2002, 114).

The in-the-moment emergence of a surf-riding identity, through the performance of the practice itself, thus becomes a compulsion that requires acting on in order for the individual to be fully realised. It is not an identity that requires articulation all the time, but when the urge comes, to not satisfy the compulsion erodes the individual's mood, balance and identity. For those that interact with water in this way, the multiple selves that come together to form them need this compulsion addressed, and these taskscapes to be articulated, in order that the in-the-moment relational sensibility can be experienced. These taskscapes and emotions become as much part of the surfing body as the muscle memories of local conditions, swells, rock walls and rips, and by turns both satiates and serves to further encourage the compulsion to surf again.

Compulsions to surf and its consequences: mobilising protection

As we have outlined in this chapter so far, there are a number of ways in which humans' relations to the surfing waters can be understood. These various conceptualisations are significant as they help us to frame and explain the ensnarement of humans by water; however, they also do more than this.

These explanations help us to understand the different ways in which aspiring surf riders act politically in the water world.

As we have seen, without waves to surf, surfers cannot be. Surfing is more to surfers than the basic premise of riding a wave; they/we get cranky if they/ we can't go; they/we sacrifice all manner of other things in order to get in the water and engage with waves, all so they/we can be surfers in the linguistically literal sense. Many surfers increasingly acknowledge that their compulsion to surf in turn creates a necessity to work politically to ensure that they and others are able to continue creating taskscapes. In this way, surfers are moving beyond accepted caricatures of disengaged beach bum, rebel or maverick (see Warshaw 2017), and evolving into politically (self-) aware citizens. As the following member of the Surfbreak Protection Society in New Zealand, and a survey respondent outlines:

> Well we were those stereotypes, right. And so yeah a lot of us, we were, we just lived to surf. And so you know [we were like that] when we were young and it's as we got older you realise that hang on here actually we're seeing these places that we value so highly, we love dearly, we want to treasure and continue to treasure it.
>
> (Surfbreak Protection Society Member 2017)

> It's the most connecting aspect to nature to me. It's the most satisfying relationship I have had with nature that I felt I did not harm. The more I connect and ride the wave, the more I feel the obligation to take care of it.
>
> (Survey Respondent 2015)

> Through the recreation of surfing I am much more aware of the issues impacting the beaches and coast. I am more aware of political decisions that impact my local breaks, much more aware of changes that happen.
>
> (Survey Respondent 2017)

As an artist would value a canvas and the potential it holds, many surfers are now working to identify and protect their invaluable materials: their coastlines and their waves. In an attempt to preserve their identity and their playground, a growing number of individuals are mobilising to create mechanisms at a variety of scales to defend their surfing spaces. The need to protect the waves is in part a result of increased pressure on littoral spaces, from development, overuse and various other coastal management issues that are threatening that 'magical mix' of features that allow waves to break. As Comer suggests, this compulsion has been driven by a necessity to secure waves from over- or mis-use by a range of humans interested in the water world:

As history has proven repeatedly, the most beloved of surf spots become subject to use and ugly overuse. Their future usability has required the invention of more responsible practices than anyone originally suspected necessary or possible. Countless green projects have sprung from this basic dilemma and insight and have politicised the rank and file of surf culture.

(Comer 2010, 205)

We can see the urge to protect to coalesce around the identity types outlined in the chapter thus far. To protect their *surf-shore identity*, surfers can channel their efforts into the preservation of particular breaks that are of importance to them. For example, 'Save Trestles' was a campaign to stop the development of a toll road that would have destroyed the famous surf spots and surrounding ecology at Trestles and San Onofre State Beach, southern California. The campaign lasted over 15 years, with surfers mobilising and engaging various groups to ultimately quash the proposals, protecting the areas and its waves into the future. As Rick Erkeneff, president of the South County Surfrider Foundation explains:

We, as a group of surfers, really grabbed ahold of this. We made it cool and sexy and we made it our fight[.] We brought in heavy hitters like the Sierra Club and other environmental groups, these die-hard environmental organizations. And we stood arm in arm and we said, 'No, we're not going to allow this.'

(Kwong, Williams and Connelly 2016)

This is not an isolated example. Surfers have acted on their surf-shore identity by leading the charge to protect a range of beaches from various threats, including development, access or water quality (Punta de Lobos, Chile and Cowells Beach, USA, for example). In addition to specific threatened breaks, surf communities are also working to protect certain regions, or clusters of surfing sites that are deemed to be of significance to local and global surf culture. World Surfing Reserves, a programme run by Californian non-governmental organisation Save the Waves, is an example of how surf communities are seeking to protect their surf-shore identity by declaring the importance of local waves on a global level. There are 10 World Surf Reserves now designated, based on a criteria of (1) quality and consistency of the wave(s); (2) environmental characteristics; (3) culture and surf history; and (4) capacity and local support. The local support, driven by this desire to continue to be able to reach that feeling of 'just right', is key to the selection and ongoing functioning of the Reserve; as Mucha states:

The local community rallied against it (a cruise ship terminal) and Andrew and other local activists organised a rally at Kirra, they made it very clear that they didn't want a cruise ship terminal they wanted a WSR [World Surf Reserve]. So here's hundreds or maybe 1,000 people saying WSR in the sand which sent a very strong message to us, in our criteria, that community support was there.

(Mucha 2017)

Beyond a connection to a specific break or group of breaks, many surfers act on their affinity to the ocean in general by mobilising around their *sea-space identity*. Organisations such as Surfers Against Sewage (SAS), for example, support site-specific activism such as beach cleans, but focus their resources on broader issues such as water quality and plastics pollution. Originally a campaign group fighting for clean waters to surf in, SAS is now one of the UKs 'most active and successful environmental charities' (Surfers Against Sewage 2017). Regarded highly by both public and politicians, SAS now serve as a parliamentary advisory body and run educational programmes in addition to their direct activism.

Being mobilised by surf-shore and sea-space identities therefore not only helps to conserve the material environments that individuals care about, but by implication it also protects the potential for these individuals to surf in them. In other words, these mobilisations secure for individuals the opportunity to act on their compulsions and emerge as a surfer. Explicit political mobilisations that seek to secure these opportunities for emergence can be identified in campaigns that aim for free access to beaches and the waves beyond. Such campaigns can be seen in the UK where surfers have waged long-running actions to secure access to Ministry of Defence coasts in Devon (Nelson, Cummins and Tagholm 2013); and in the US where private ownership threatens access in areas of California (Davidson 2005). In Australia, chapters of NGO Surfrider Foundation are striving to help less mobile people, surfers or not, have access to beaches, by 'strongly encouraging' authorities to provide and promote 'barrier-free' access to beaches and public coastal lands (Surfrider Australia 2015).

Interestingly, however, new political actions are occurring that allow individuals to emerge as surfers in different ways. Wave pools, where rideable artificial waves are offered to surfers, are now becoming a feature in the surfing world. Operating commercially for public use, and also by professionals in a competitive context (see Borne and Ponting 2017), a range of systems are now in existence, offering their version of the perfect wave at the flick of a switch (see Figure 6.4). While a number of parks are planned to serve areas where surfing is already very popular, such as the Gold Coast, Australia, others have been developed in locations far from the nearest coastal surf spot, such as in Austin, Texas (Surfpark Central 2017).

While these are not perhaps political moves by surfers in a true sense, they have an ambivalent relation to the politicised issue of overuse and present a

Figure 6.4 Catching a wave at Surf Snowdonia, a wave pool in Wales (L. Stoodley 2015).

range of questions for surfers and researchers of human–water relations. Do wave pools hold the potential to translocate surfers from one area to another, and thus solve problems of overcapacity at particular local breaks? Or, will wave pools generate greater interest in surfing and exacerbate crowds at all locations? As technology improves and access to artificial waves becomes easier, will the surfer be able to resist the compulsion to create that performative taskscape however it is generated, and wherever it is? Can the stoke created by riding your local wave, or the energy in the ocean, by re-placed by waves generated at the push of button?

Conclusion

This chapter has explored how many humans have come to define their sense of self through their activities and practices in the surf zone. When conditions align in these unique liquid scapes, the elements combine to create a 'liquid gold', breaking waves that are sought out and ridden by self-identified water people, on craft, board or body. This chapter has argued that surfers develop a connection to the sea and surf through regular engagement, and that many surfers become so attracted to the alchemic make up of these surf zones that they experience a *compulsion* to participate in this practice. This is a phenomenon that not only establishes surfing as an essential aspect of the surfers' life, but also as a key element in the development of their identity. It is in these spaces that the surfers feel most at home, when they are able to *be* surfers. At other times, surfers recall previous surfing experiences, until the next swell arrives and they allow themselves once more to obey 'dog-whistle orders from the collective surf unconscious' (Finnegan 2015, 170), and reengage with their favoured space. Suggesting that such a compulsion can be compared to the work of a sculptor or painter, the chapter proposed that surfing can be viewed as an art form, with each wave representing a new canvas for the generation of split-second masterpieces. Many surfers would argue that these works are worth protecting, and the final section of this chapter has discussed how political action and environmentalism has found its place in the surf lifestyle. The development of surfer-led initiatives has provided surfers with a means through which they can enter new arenas of political participation, to portray the importance of certain surf zones to influential, wider audiences. While surfers appear to have an inherent understanding of this importance, attempts to explain this to those unaffected by the magnetic appeal of the littoral zone have in the past been ineffective, casting surfers to societal fringes. In taking this new approach, surfers are positioning themselves within the system, creating an 'enviro-surf' agenda to protect their most valued surfing spaces and allow for compulsive calls to the sea to be fulfilled into the future.

References

Anderson, J. (2004) Spatial Politics in Practice: The Style and Substance of Environmental Direct Action. *Antipode*, 36(1), 106–125.

Anderson, J. (2009) Transient Convergence and Relational Sensibility: Beyond the Modern Constitution of Nature. *Emotion, Space, and Society*, 2, 120–127.

Anderson, J. (2014a) Surfing Between the Local and Trans-Local: Identifying Spatial Divisions in Surfing Practice. *Transactions of the Institute of British Geographers*, 39(2), 237–249.

Anderson, J. (2014b) What I Think About When I Think About Kayaking. In J. Anderson and K.E. Peters (eds), *Water Worlds: Human Geographies of the Ocean*. Farnham: Ashgate, pp. 103–117.

Anderson, J. (2015a) On Being Shaped by Surfing: Experiencing the World of the Littoral Zone. In M. Brown and B.E. Humberstone, *Seascapes: Shaped by the Sea*. Farnham: Ashgate, pp. 55–70.

Anderson, J. (2015b) *Understanding Cultural Geography: Places and Traces*. London: Routledge.

Anderson, J. and Peters, K.E. (2014) *Water Worlds: Human Geographies of the Ocean*. London: Routledge.

BBC. (2017, 5 December) *BBC Weather – Porthcawl*. Retrieved from www.bbc.co.uk/weather/2640054

Beljaars, D. (forthcoming) Towards Compulsive Geographies. *Transactions of the Institute of British Geographers*.

Beljaars, D. and Anderson, J. (forthcoming) Tourette Syndrome through the Eye of the Beholder. *Geohumanities.*

Borne, G. and Ponting, J. (eds). (2017) *Sustainable Surfing.* Abingdon: Routledge.

Brooks, S. (2017, 19 May) *Love Thy Local – Maroubra.* Retrieved from www.tracksmag. com.au/news/love-thy-local-8211-maroubra-462301

Brown M. and Humberstone, B.E. (2015) *Seascapes: Shaped by the Sea.* Farnham: Ashgate.

Butler, J. (1990) *Gender Trouble: Feminism and the Subversion of Identity.* London: Routledge.

Butler, J. (1993) *Bodies that Matter: On the Discursive Limits of 'Sex'.* London: Routledge.

Capp, F. (2003) *That Oceanic Feeling.* London: Aurum.

Casey, E. (2001) Between Geography and Philosophy: What Does It Mean to Be in the Place-World? *Annals of the Association of American Geographers,* 91(4), 683–693.

Chamberlain, B. (1996) *Tide Race.* Bridgend: Seren.

Comer, K. (2010) *Surfer Girls in the New World Order* (1st ed.). Durham and London: Duke University Press.

Crang, M. and Travlou, P.S. (2001) The City and Topologies of Memory. *Environment & Planning D: Society & Space,* 19, 161–177.

Davidson, R.A. (2005) The Los Angeles Coast as Public Place. *Geographical Review,* 95 (4), 578–593.

Ellis, C. (2004) *The Ethnographic I: A Methodological Novel about Ethnography.* Walnut Creek, CA: AltaMira Press.

Evers, C. (2009) 'The Point': Surfing, Geography and a Sensual Life of Men and Masculinity on the Gold Coast, Australia. *Social and Cultural Geography,* 10(8), 893–908.

Evers, C. (2010) *Notes for a Young Surfer.* Carlton: Melbourne University Press.

Featherstone, M. (1995) *Undoing Culture.* London: Sage.

Finnegan, W. (2015) *Barbarian Days.* London: Corsair.

Goffman, E. (1977) *Frame Analysis: An Essay on the Organisation of Experience.* Cambridge, MA: Harvard University Press.

Halbwachs, M. (1992) *On Collective Memory.* Chicago: University of Chicago Press.

Haraway, D. (2003) *The Companion Species Manifesto: Dogs, People, and Significant Otherness.* Bristol: Prickly Paradigm.

Heath, S. (1981) *Questions of Cinema.* Basingstoke: Macmillan.

Heller, P. (2010) *Kook: What Surfing Taught Me About Love, Life, and Catching the Perfect Wave.* New York: Simon & Schuster.

Holt, L. (2008) Embodied Social Capital and Geographic Perspectives: Performing the Habitus. *Progress in Human Geography,* 32(2), 227–246.

Ingold, T. (1993) The Temporality of the Landscape. *World Archaeology,* 25(2), 152–174.

Jameson, F. (1991) *Postmodernism or the Cultural Logic of Late Capitalism.* London: Verso.

Kotler, S. (2006) *West of Jesus: Surfing, Science, and the Origins of Belief.* New York: Bloomsbury.

Kwong, J., Williams, L. and Connelly, L. (2016, 11 November) *After a 15-Year Battle, Trestles Surf Spot Is Saved.* Retrieved from the Orange County Register: www. ocregister.com/2016/11/11/after-a-15-year-battle-trestles-surf-spot-is-saved/

Leza, J. (2012) *Diaries of a Surf Widow.* Retrieved from www.wavescape.co.za/blog/ the-surf-widow-diaries/lessons-learned.html

Magicseaweed. (2017, 5 December) *Porthcawl – Rest Bay Surf Report and Forecast*. Retrieved from https://magicseaweed.com/Porthcawl-Rest-Bay-Surf-Report/1449/

Mattos, B. (2004) *Kayak Surfing*. Bangor: Pesda Press.

Surfbreak Protection Society Member. (2017, 3 December). Interview with Lyndsey Stoodley.

Mucha, N. (2017) *World Surfing Reserves: Protecting Iconic Surf Environments*. Gold Coast: International Surfing Symposium.

Nash, C. (2000) Performativity in Practice: Some Recent Work in Cultural Geography. *Progress in Human Geography*, 24(4), 653–664.

Nelson, C., Cummins, A. and Tagholm, H. (2013) Paradise Lost: Threatened Waves and the Need for Global Surf Protection. *Journal of Coastal Research*, 65, 904–908.

Olive, R. (2016) Going Surfing/Doing Research: Learning How to Negotiate Cultural Politics from Women Who Surf. *Continuum*, 30(2), 171–182.

Peters, K., Stratford, E. and Steinberg, P. E. (2017). *Territory Beyond Terra*. Rowman & Littlefield.

Pratt, M. (2000) The Good, the Bad, and the Ambivalent: Managing Identification Among Amway Distributor. *Administrative Science Quarterly*, 45, 456–493.

Rasmussen, S.A. and Eisen, J.A. (1992) The Epidemiology and Clinical Features of Obsessive Compulsive Disorder. *Psychiatric Clinics of North America*, 15, 743–758.

Reilly, D. (2003) *Only a Surfer Knows the Feeling: The Story of Billabongs Surfwear Revolution*. Bondi: Rolling Youth Press.

Richards, M. (2002, December–January) Real Surfing: Agree to Disagree. *The Surfers Path*, 28, 114.

Ricketts, E., Woods, D., Antinoro, D. and Franklin, M. (2012) Phenomenology and Epidemiology of Tic Disorders and Trichotillomania. In G. Steketee (ed.), *The Oxford Handbook of Obsessive Compulsive and Spectrum Disorders*. New York: Oxford University Press, pp. 89–107.

Sack, R. (1997) *Homo Geographicus: A Framework for Action, Awareness and Moral Concern*. Baltimore: John Hopkins University.

Stoodley, L. (2017, June) Reflective journal.

Surfers Against Sewage. (2017, 24 November) *About Us*. Retrieved from www.sas.org.uk/about-us/.

Surfpark Central. (2017, 27 November) www.surfparkcentral.com/.

Surfrider Australia. (2015) *National Campaign Policy on Beach and Foreshore Access*. Retrieved from www.surfrider.org.au/campaign_policies.

Survey Respondent. (2015, 1 July) Surf Survey: Purpose built and Artificial Waves.

Survey Respondent. (2017) Surf Survey: World Surfing Reserves.

Sutton-Smith, B. (1997) *The Ambiguity of Play*. Cambridge, MA: Harvard University Press.

Tess. (2011, 24 October) *Life Is Better When You Surf*. Retrieved from Global Industries Blog. Retrieved from https://blog.surfindustries.com/2011/10/life-is-better-when-you-surf/.

Turnbull, D. (2002) Performance and Narrative, Bodies and Movement in the Construction of Places and Objects. *Spaces and Knowledges*, 19(5–6), 125–143.

Usher, L. (2015) 'Foreign Locals': Transnationalism, Expatriates, and Surfer Identity in Costa Rica. *Journal of Sport and Social Issues*, 39(6), 455–479.

Vannini, P. (2015) Non-Representational Ethnography: New Ways of Animating Lifeworlds. *Cultural Geographies*, 22(2), 317–327.

Warshaw, M. (2017, 27 June) *Mike Hynson: Rebel Rebel*. Retrieved from www.surfer.com/blogs/eos/mike-hynson-rebel-rebel/

Williams, R. (1977) *Marxism and Literature*. Oxford: Oxford University Press.

Wood, N. (2012) Playing with 'Scottishness': Musical Performance, Non-Representational Thinking and the 'Doings' of National Identity. *Cultural Geographies*, 19(2), 195–215.

Zink, R. (2015) Sailing Across the Cook Strait. In M. Brown and B. Humberstone (eds), *Seascapes: Shaped by the Sea*. Farnham: Ashgate, pp. 71–100.

Chapter 7

Waves as emblemata for knowledge

John Hartley

This chapter looks at two examples of the author's arts practice alongside the sea, towards which they are directed, as forms of knowledge-making. It starts with an initial description of the complex and multi-layered nature of waves in real-life situations, before discussing how art processes are also constituted through shifting connections of greater or lesser prominence. This allows a discussion of the shapes or forms of knowledge-producing practices and the facts produced by those processes; something Lorraine Daston has described as the *emblemata* of knowledge. Daston challenges the characterisation of 'modern' facts, as 'Rocks: hard, jagged, plain rocks … the thugs of epistemology' (Daston cited in Latour and Weibe, 680–681: 2005). In reality, facts are often constructed through negotiation and in conditions of uncertainty, including social processes and contingent equipment; 'Far from being obdurate, much less obstructionist, facts are often faint and flickering … As the emblemata of facts, rocks are relative newcomers' (ibid.).

Consideration of the two art processes I use (first drawing and second hacking of objects) and their appropriateness for addressing the sea highlights the conditions that help shape the artwork. These include material, psychological and societal processes, entangling the object of study and the subject that is undertaking that study. Taken together (as what I term the art apparatus) these social and material constituents display emergence and collapse in ways that are familiar from the intensive changes within the sea. Comparisons between the sea and knowledge-making about it present waves as alternative emblemata for various forms of knowledge than the distinct and obdurate rocks Daston relates to the knowledge of a modern period.

Sea knowledge

The changes of the sea are constant and varied. Waves in the sea present an infinite number of closely related shapes, distributed over many different scales. The largest sea waves are vanishingly slow, determined by continent formation and driven by oscillations of heavenly bodies. Long waves are generated by fluctuations in the Earth's crust and atmosphere. Tides take place over hours and have a wavelength of hundreds, to thousands, of kilometres (Holthuijsen 2010, 3–5).

The forms of the waves themselves are influenced by the depth of the water and shape of the seabed, the interplay of currents and rebounding backwash resonating off masses they encounter and, in their turn, change. Surges caused by the low pressure of storms (patterns of rising and falling atmospheric pressure) take place over a few hundred kilometres and a few days. Other oscillations include unpredictable *seiches;* standing waves that develop within partially bounded 'resonant basins' (such as a harbour or bay). More noticeable on a human scale are waves generated by wind, and the water's own earthly gravity. Ripples (which oceanographers refer to as *capillary waves)* ride on and confuse the pattern of these planetary volumes of water sloshing around a bumpy and irregular surface. *Surface gravity waves, wind sea* and *swell* occupy temporal niches between ¼ and 30 seconds in periodicity. Groups of these waves generate *surf beat* over a few minutes (Holthuijsen 2010, 3–5), yet another set of patterns nested within an array of oscillations over many scales.

Intensity

These wave figures do not describe the material of the sea, but rather the behaviour of that material. They are a figure of change. Looking carefully at the movement of the waves, water circulates within the wave but does not itself travel with the wave (Wiegel 1964, 58). Waves are a form of propagating change; a 'signal' of energetic difference that is not clear-cut or easily isolated.

Manuel De Landa (2004) uses the term intensities when talking about complex material change. De Landa differentiates between systems that can be measured and touched – extensive systems – and those that cannot, which he calls intensive. If we view the movement of waves from a different perspective to the movement of the water particles, we can think of waves as an 'intensive' movement or communication. They are an energetic signal that travels through the extensive saltwater medium (as well as other substrates such as atmosphere and geological formations).

Drawing waves

I undertook a series of drawings that aimed to explore how artistic skills could register the movements and change that animate the sea and so articulate some of its connected, intensive nature. These drawings were informed by the physical experience of being in and moving among waves and were undertaken following sea swimming (Figure 7.1).

A swimmer's body rolls and pitches on different axes while in the water. It bobs at a frequency influenced by its mass, density and arrangement, oscillating around a centre of balance that is continually changing through its own cycles of a swimming stroke. While swimming, arms 'catch' and 'pull' in

Figure 7.1 Sea-swimming drawing (2016), ink on paper.

cycles with legs kicking at a half or a third of that frequency. These cycles are taken into the waves and meet with, combine and subtract from them. The swimmer is slowed down or sped up as waves pass. Exhaling slowly over the time it takes to make three strokes, the swimmer's lungs empty of air and their torso decreases buoyancy. As a result, the centre of balance moves and the head sinks lower. After perhaps three strokes the head turns and air is inhaled. The top half of the body becomes more buoyant, and tips up in the water again. All of this occurs amidst the movements of the sea, also changing the waves on a local scale and producing new ripples and splashes. The resultant movements can be considered as oscillations in multiple axes and within systems of different scale (in time and space). They produce a complex, changing engagement that is not easily separated into actions of sea and actions of swimmer. The process of swimming in the sea is affected by and changes the waves in which it takes place. The effects of swimmer and sea are entangled (Figure 7.2).

The drawings were undertaken using a process of line repetition, where new lines responded to existing ones, copy their character, noticing and repeating any variations and allowing the drawing to progress or emerge in unexpected ways as it develops. This was a system of sorts, albeit a rather loose, or fluid one.

Figure 7.2 Sea-swimming drawing (2016), ink on paper.

The completed drawings manage to show something at least, of the complex oscillations that are experienced in the sea. Just as sea conditions are extremely variable, so the drawings move between very subtle, almost invisible series of marks, to the portrayal of much starker events. They exhibit a sense of eddy and rhythm that doesn't repeat exactly, but progresses and changes over varying periods (whether from line to line, or between areas of the drawing). There is a sense of movement and change existing beyond the frame of the drawing, or outside its plane of visibility, suggesting if not bottomlessness, then at least significant continuation beyond the frame of immediate perception. They address conditions that are not easily bounded. A sense of oscillations of multiple scale are suggested by the drawings. Transitions (like those between rock and sand, flat calm and rough) have been brought about by a slight push in the elastic responsiveness of a line to its predecessor (as shown in Figure 7.3).

This emergent process responds to and changes moment by moment based on the 'events' that occur within the drawing. Using a mix of recollection, research, artistic training and continued attention, the artistic process recalls or evokes a physical experiences but it is also reminiscent of other forms, including a range of observed natural conditions or visualisation of other systems.

Here I am describing the drawing process as if the artist were not present (and in common with much art relating to landscape traditions, there is no central figure or subject), just a field of conditions. When used as a way of producing knowledge about conditions in the sea, this sort of art process appears with much of its workings hidden. The process almost suggests the drawn line itself were making the decisions and was the agent of its own manifestation. Or as if the drawing was derived directly from the sea; a trace of objective material conditions.

Of course this is not the case. The process of concentrating on the emerging character of line brought a range of subjective tools to bear within the drawing. Novelty, boredom, a sense of tidiness, clutter, or impatience were mobilised in ways that are not easily managed or measured. Micro-

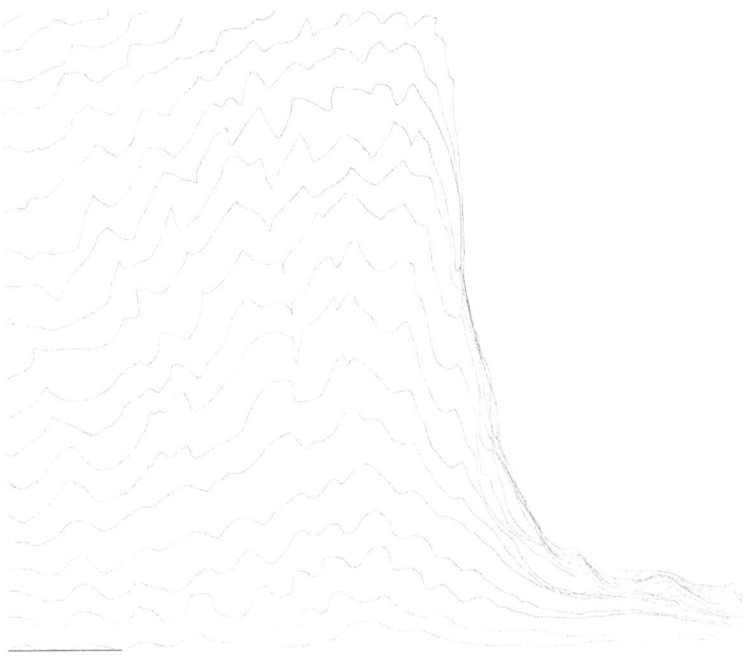

Figure 7.3 Sea-swimming drawing (2013), ink on paper.

experiences position the human artist as an inescapable part of this partially controlled, yet largely open system (a drawing), which aspires to register in some way (using memory, skill and playful speculation) aspects of other systems beyond the process of drawing alone. Furthermore, one reason for addressing the sea within this process was that the works should draw attention to wider systems and processes, relevant to ecological change.

Process as apparatus

Many other processes are working (and oscillating) during such an immersive act of research. Having experienced an initial shock from the cold, the swimmer's mammalian dive response reduces heart rate and constricts peripheral blood vessels (Bruce and Speck 1978, 9). Exercise increases metabolism and generates heat in muscles, another change working in a different 'direction'. Lactic acid builds up and body temperature changes slowly as blood flows away from the extremities to the core of the swimmer's body. Breathing and stroke are affected by these temperatures and exertions.

The iterations and oscillations the swimmer experiences exist in other substrates than sea water and other systems of greater complexity. A swim may start from a beach of finely sifted sand accumulated by the predominant, ocean-dominated swell. Such sand is often collected in bays of collapsed rock, eroded over countless storms. Geological oscillations meet cycles of marine movement to configure the field in question. Moving into deeper water increases the 'chop' of the sea. Small-scale movements of the body and arms now have to respond differently to the depth below, affected by more distant wave patterns and deeper slopes of sediment falling away from land and slowly sinking towards the edge of the marine shelf. Returning to the beach, a swim route may be disrupted, involving a struggle to pass through bands of floating kelp, which, having grown following its own circannual rhythms influenced by day length (tom Dieck 1991, 341–350), is often torn loose by prior storms the frequency of which are also changing over recent years (Met Office 2014). Other levels of oscillation are overlaid in very different, but co-present systems. This activity passes through and registers many layers of interacting oscillation. Waves in metabolism, geology, algal cycles and physical movement stack on top of each other.

Considering the interconnections of the drawing process encourages an awareness of how human actions and systems are combined with non-human processes to produce the works. This asks new things of the art process, requiring a different approach or materials that could increase the visibility of some of the infrastructures and conditions within their outcomes, and as a consequence locate the human actors differently within this ever-changing ecosphere.

Karen Barad's method for exploring the outside boundary of an 'apparatus' (which builds on Foucault's treatment of apparatus or dispositif[1]) traces the entanglements of material conditions with subject positions (Barad 2007, 223). Applying this method to art that addresses the conditions of the sea can help describe a knowledge-producing apparatus that addresses material conditions, while also being a psycho-social practice dependent upon and responding to, conditions that are easily overlooked.

And innocuous as it is, the process described above is indeed part of a wider system, parts of which are unsustainable or heavily resource-dependent in a way that could have implications for marine ecologies, be it at a remove. Many sea swimmers use neoprene wetsuits, silicone hats and polycarbonate goggles. This equipment is produced from oil and manufactured and traded with a particular form of globalised economics. A swimmer's choice of beach may benefit from municipal maintenance and be chosen for its access to transport infrastructure. To warm up after an exhilarating swim, sea-swimmers often enjoy a hot drink from a local beach cafe; made from coffee or cocoa beans from Colombia or Kenya. But they many not think of their connections to and dependencies on many infrastructures of government, trade and manufacture while basking in the rush of endorphins that accompany cold-water swimming (Outdoor Swimming Society 2012). As well as physical and material processes a swim in a 'natural' location depends upon and changes many areas of life, be they local or distant, social, political, economic or physiological.[2] There are many entanglements and interferences between the conditions and the process used to observe them. But few are acknowledged or visible in the drawings.

These drawings are not an objective picture of a non-human system (the sea) in isolation, but rather derive from a combination of systems, including the sea, but also the physical and metabolic actions of a swimmer (relayed through human memory), the specialist expertise of the drawing practitioner and the fleeting psychological static of concentration. These multiple systems together operate as a form of apparatus. A distributed and entangled series of effects are enacted along that apparatus (ultimately resulting in traces on a piece of paper).

This evident subjectivity appears to undercut the drawings' ability to make anything but the loosest comments about the objective material conditions to which the drawing process was directed, if we are seeking to observe those conditions in isolation. But considering the drawing process more carefully starts to show how subjectivity can be brought into light alongside the material changes of non-human processes. They suggest an art apparatus (responding to subject and object conditions) might be able to show something of material conditions around the sea and some of the different sorts of entanglements that involve those looking at or towards the sea.

Hacking waves

In order to move beyond the limitations of the drawings, I produced other artworks that attempted to register the art apparatus more clearly while also seeking to reduce its dependence on unsustainable resource use. I used an approach based upon hacking (or to use Claude Levi-Strauss' term bricolage[3]), which reused existing materials and equipment (such as discarded electronic consumer products or domestic objects) that were broken down and combined in novel ways to give rise to new arrangements and objects.[4] By highlighting the resource implications of their own production and use, this made some of the processes with which the apparatus was connected more visible than with the drawing process described above. Resulting in a series of artworks (or *Hydrohacks*) that could be taken directly into the sea, the works reused materials such as personal stereo headphones and other common electrical components to make low-cost hydrophones; and for another work reshaped the material resources at the heart of an office (suits, a desk and a computer), which were broken down and reformed as a functioning sea kayak. Additionally, a series of underwater video cameras were produced from old and redundant DV cameras that had lost much of their commercial value. These were waterproofed in pickle jars and sandwich boxes, and here I will focus on these underwater swimming cameras in comparison to the sea-swimming drawings discussed above.

One work involved waterproofing a Mustek DV2000 video camera in a domestic pickling jar (a classic tool of self-sufficiency and literal preservation, familiar to American survivalists and suburban allotment holders alike). Together with improvised means of tying the jar to a swimmer's torso, this apparatus was capable of making films in shallow underwater depths. The reused camera was initially manufactured around the year 2000 at which point it retailed for above £100 and its 3-megapixel specification was a key selling point. The camera used in this artwork was found on a second-hand website for a very low price. It offered exactly the same functionality as it did when it was cutting edge, in the year 2000 (Figure 7.4).

Despite the decline in its commercial value, the equipment worked just as well as it ever did. Nowadays it is regarded as well behind the *innovation adoption curve* (Rogers 1962) but the technology has not changed, only conditions around the technology (Figure 7.5).

Figure 7.4 Swim camera (2012), redundant DV cam, pickle jar.

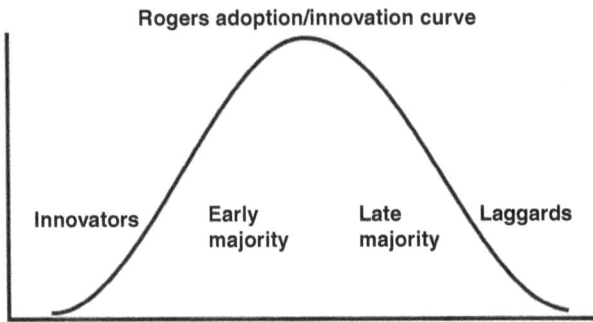

Figure 7.5 Visualisation of change in social adoption of innovations (after Rogers 1962).

This curve can be read as a wave in a socio-techno substrate. It describes changes in production norms, expectations of specification and penetration of a particular market. Changes in the conditions of production and adoption for consumer electronics (such as the camera I used) could be described as intensive changes that affect otherwise unchanged (extensive) components. Technical options, factory capacity, supply lines and future construction options peak and decline variously, so do consumer preferences, expectations and retail price. As the wave of innovation progresses, the redundant object is not carried 'forwards' with it. Components sink down into social neglect; perhaps more or less slowly depending on whether the product enjoys a 'long tail' of continued use (Anderson 2006). This change in relations between society and material technologies is analogous to the movement of water particles – rising and sinking – animated by a waveform that emerges and collapses. And this technological wave is (as with water) intensive. The equipment doesn't necessarily change. It is our psychological reception of it that changes as well as the wider social and material conditions required for its use including other devices, software, manufacturing chains that it 'plugs into'.

Films made with this equipment give a strong impression of sea swimming. Bubbles and limbs pass in ways that are calming then disorientating by turns. The movement of the swimmer, the sound of the water, the strokes of arm and rotation of the body are all evident. As with the drawings, multiple waves and oscillations of extensive movement combine on top of each other. The waves' substrates of water and sand are visible as clear, deep light and colours. And the intensive waves of technological adaptation and its social dissemination are also present in the produced video. The grey-greens of the deep, the intense whites of the sky and the diagonal cast of the movement and sea, and through the camera's imperfect colour responses, its pixilation and its state of outmoded glitch. Additionally these intensive and distributed waves are highlighted by the hacked camera itself; an art object that draws attention to infrastructures from which it came and their own capacity for change. In this case, the tools of investigation reveal something of their own conditions and processes and how they share ways of changing with the objects to which they are directed.

Waves as emblemata

Looking at two different art processes attempting to register the conditions of the sea has highlighted types of movement and change in contexts that were both human and non-human. These included complex spatial patterns in sea movement derived from interaction between swimmer, sea and wider rhythms of geophysics. By overlaying a process of disciplinary training, equipment, attention and recollection drawings were produced that registered aspects of these systems, in semi-controlled or unexpected ways. Subsequently, hacking recalled the geometries of change shown in the previous drawings, but showed them within the art processes and the infrastructures they depended upon. In this way, foregrounding the tools and processes of the art apparatus led to a series of works that registered conditions of the sea (its colour, light and movement), but also presented the changes of the art apparatus itself in a way that could be considered as sharing something of the movements of the sea. Intensive waves were traced in art and in the sea, so suggesting different possibilities for thinking about the material relationships that are part of knowledge-generating practices.

Given the distributed, multi-scale nature of change within the Anthropocene, greater attention to the context of production (as explored by the *Hydrohacks*) asks questions that could also be directed towards other knowledge-making practices. Artistic attention to the context of production was able to increase the visibility of the tools and conditions upon which the findings of the process were dependent. The works highlight how they are implicated in the system they seek to look at and how their knowledge production registers (to differing degrees of clarity) material objects of investigation as well as multiple subject positions including individual or societal process and infrastructure within the apparatus. These art processes have registered something of the behaviour of intensive, interconnected waves, but not in ways that could be considered repeatable and objective. They haven't presented something that they sit apart from, but rather something that they are part of – complex movement and change that entangle the conditions under investigation with processes of knowing.

Notes

1 See Agamben (2009).
2 I develop this second description of a sea swim in a similar manner to Barad's sketch of the multiple connections and events triggered by a traveller working on a plane (Barad 2007, 223).
3 Claude Levi-Strauss contrasts what he calls the 'bricoleur' with the engineer. While the engineer has a rational and planned relationship between task, available material

and planned project aim, the bricoleur has to make do with available resources, which may not relate to '[t]he current project, or indeed to any project [but are] the contingent result of all the occasions there have been to renew or enrich the stock or to maintain it with the remains of previous constructions or destructions' (Levi-Strauss 1966, 17).
4 By addressing the resources used for their construction, these works consider their own environmental presence and the possibility that they are, to some degree, implicated in the conditions that they are attempting to address. Their low financial and environmental cost (due to reuse of resources) also contributed to this knowledge-making approach being more socially accessible. The approach builds on methods developed within tactical media arts practice such as the Zero Dollar Laptop (ZDL 2014), or the Redundant Technology Initiative (RTI 1996), which address some of the social, creative and environmental significance of reusing hardware.

Bibliography

Agamben, G. (2009). *What Is an Apparatus?* Stanford: Stanford University Press.

Anderson, C. (2006). *The Long Tail: Why the Future of Business Is Selling Less of More.* New York: Hyperion.

Barad, K. (2007). *Meeting the Universe Halfway: Quantum Physics and the Entanglement of Matter and Meaning.* Durham, NC: Duke University Press.

Beck, U. (1992). *Risk Society: Towards a New Modernity.* Trans. M. Ritter. New York: Sage.

Beech, D. (2006). Institutionalisation for All. *Art Monthly* 294(March): 7–10.

Bohm, D. and Peat, D. (2000). *Science, Order, and Creativity.* London: Routledge.

Bruce D. and Speck, D. (1978). Effects of Varying Thermal and Apneic Conditions on the Human Diving Reflex. *Undersea Biomedical Research* 5(1):9–14.

Brundtland Commission. (1987). *Our Common Future.* New York: United Nations.

Bryant, L. (2010). Larval Subjects: Hyperobjects and OOO. Accessed 11 November 2010. https://larvalsubjects.wordpress.com/2010/11/11/hyperobjects-and-ooo/.

Crutzen, P. (2002). Geology of Mankind. *Nature*, 415(6867), 23–23. doi:10.1038/415023a.

De Landa, M. (2004). *Intensive Science and Virtual Philosophy.* London: Continuum.

Deleuze, G. (2006). *The Fold.* London: Continuum.

Deleuze, G. and Guattari, F. (1980). *A Thousand Plateaus* (English Translation). London: Continuum.

Ellul, J. (2006). *The Technological Society.* New York: Vintage.

Gunderson, L. and Holling, C. (2002). *Panarchy: Understanding Transformations in Human and Natural Systems.* Washington, DC: Island Press.

Haraway, D. (1992). *The Promises of Monsters: A Regenerative Politics for Inappropriate/D Others.* New York: Routledge.

Hartley, J. (2014). Turn Your Office into a Kayak. Accessed 5 June 2014. www.instructables.com/id/Turn-your-office-into-a-kayak/

Holthuijsen, Leo H. (2010). *Waves in Oceanic and Coastal Waters.* Cambridge: Cambridge University Press.

Latour, B. (1993). *We Have Never Been Modern.* Trans. C. Porter. Cambridge: Harvard University Press.

Latour, B. (2009). *Politics of Nature.* Cambridge: Harvard University Press.

Latour, B. and Weibel, P. (2005). *Making Things Public*. Cambridge: MIT Press.

Levi-Strauss, C. (1966). *The Savage Mind*. Chicago: University of Chicago Press.

Met Office. (2014). Wettest Winter for England and Wales Since 1766. Accessed 26 March 2014. www.metoffice.gov.uk/news/releases/archive/2014/early-winter-stats.

Morton, T. (2014). *Hyperobjects*. Minneapolis: University Of Minnesota Press.

Outdoor Swimming Society. (2012). Accessed 15 August 2012. www.outdoorswim mingsociety.com/.

Oxford Dictionary of English, 2nd edition, revised. (2009). Oxford: Oxford University Press.

Rogers, E. (1962). *Diffusion of Innovations*. Glencoe: Free Press.

Res Cogitans. (2014). Like Water in Water: The Nihilism of 'Why', the Immanence of 'Because'. Accessed 29 January 2014. http://commons.pacificu.edu/cgi/viewcon tent.cgi?article=1057&context=rescogitans.

RTI. (1996). Redundant Technology Initiative. Accessed 1 October 2013. http://rti. lowtech.org/intro.

Shapin, S. and Schaffer, S. (1985). *Leviathan and the Air-Pump: Hobbes, Boyle, and the Experimental Life*. Princeton: Princeton University Press.

Thompson, N. (2015.) *Seeing Power: Art and Activism in the 21st Century*. Brooklyn: First Melville House.

Tom Dieck, Inka. (1991). Circannual Growth Rhythm and Photoperiodic Sorus Induction in the Kelp Laminaria Setchellii (Phaeophyta)1. *Journal of Phycology* 27(3), 341–350. doi:10.1111/j.0022-3646.1991.00341.x.

Wiegel, R. (1964). *Oceanographical Engineering*. New York: Dover Publications.

Zero Dollar Laptop. (2014). ZDL Manifesto. Accessed 17 March 2014. http://zerodol larlaptop.org/wiki/doku.php?id=zdlt:manifesto.

Part III

Water we know?

Chapter 8

Re-envisioning the hydro cycle

The hydrosocial spiral as a participatory toolbox for water education and management

Rebecca L. Farnum, Ruth Macdougall and Charlie Thompson

Introduction

Earth and biological sciences conceptualise any number of processes as cycles. Children are taught from an early age various organisms' 'life cycles', even as it can be argued that individual organism's lives are chronologically linear and populations far more interconnected than a unidirectional circle. Elements, too, are understood and taught as cycles: Literature frequently refers to multiple cycles such as carbon, nutrient, ozone, nitrogen, and environmental.

Water is no exception. The 'water cycle' or 'hydrologic (hydro) cycle' is a well-known name and concept. The basic pattern of precipitation, evaporation, and condensation forms the basis of how most people comprehend water. Yet these basic biophysical processes are far more complicated. The systems we have individually categorised influence each other in myriad ways (Linton et al. 2014; McLaughlin et al. 2007). This chapter reviews how the 'classic' hydro cycle was created in Western hydrological sciences, considers hydrosocial relations literature questioning this epistemology of water, introduces the hydrosocial spiral as a new tool for understanding the movement of water, and provides preliminary results from its application. An interdisciplinary team of authors made up of an anthropologist, an artist, and a hydrologist emphasise the dynamic, iterative approach needed to understand water and its movements in our world.

The 'classic' hydro cycle: a limited approach to water issues

Jamie Linton's *What Is Water?* (2010) incorporates an extensive review of the history of the hydrologic cycle, providing a helpful distinction between the origins of hydrology as a scientific field and the creation of the diagrammatic depiction of the hydro cycle. The hydro-cycle diagram as we think of it today is a modern creation less than 100 years old, but humans have been considering water for millennia, often in a cyclical fashion. The Hebrew book of Ecclesiastes 1:7 (quoted in Leopold 1960) alludes to returning flows; similar ideas can be seen in the works of Greek and Roman philosophers and poets (Brutsaert 2012).

Pierre Perrault can be considered the world's first 'pure' hydrologist. His 1674 book, *On the Origin of Springs*, sought to describe and calculate water in its different forms. Brickner continued this work in 1905, attempting to quantify global water resources (UNESCO 1971). Around this time, hydrology became more clearly defined as an academic discipline. A 1931 paper by Robert E. Horton, 'The Field, Scope and Status of the Science of Hydrology', established the parameters for the science of hydrology. The paper included the first notable diagram depicting water movements and an argument for separating the natural and social aspects of water, sparking discourse around a diagrammatic representation of biophysical water flows. Diagrams varied extensively for the first decade, as hydrologists considered various potential depictions. Eventually, Thorndike Saville coined the term 'hydrologic cycle' to describe the depiction referred to here as the 'classic' hydro cycle (Linton 2010). The US National Resources Board published early versions of this diagram in 1934 (Linton 2008), with Mienzer's (1942) textbook including similar ideas.

Today's hydro-cycle diagrams look much different than Horton's first depiction in 1931, but surprisingly similar to the one published by the National Resources Board in 1934. Visual changes are more numerous than conceptual alterations: The diagrams found in an average primary school textbook currently look more realistic, but continue to show arrows representing unidirectional flows of water through almost entirely biophysical processes. Compare the two images in Figure 8.1: the US National Resources Board's 1934 version and the US Geological Survey's current conception. The graphics are improved aesthetically and more processes are now shown, but the two are incredibly close theoretically, especially given the nearly 80 years of modern scientific research separating them. The current hydro cycle depicted by the US Geological Survey can thus be considered the 'classic' hydro cycle.

Since the 1934 publication of the US National Resources Board's hydro-cycle diagram, various versions of the classic hydro cycle have been used in teaching natural sciences. There has been relatively little research on students' knowledge and perspectives of the hydro cycle (Shepardson et al. 2009), though it is taught in most classrooms as part of science curriculums, and so widely accepted that a basic understanding of the classic cycle is necessary for what Cockerill refers to as 'water literacy' (2010, 151). Cockerill posits that not understanding the basics of hydrology (e.g. believing that the planet could literally run out of water) decreases people's willingness and capacity to take action or change habits around water use. Cockerill ran a program in North Carolina 'translating' hydrology and scientific knowledge to make it accessible for a general audience, reinforcing the classic hydro cycle.

Nor is this emphasis limited to Western classrooms. Taiwo et al. (1999) explored how the hydro cycle is understood by schoolchildren in Botswana,

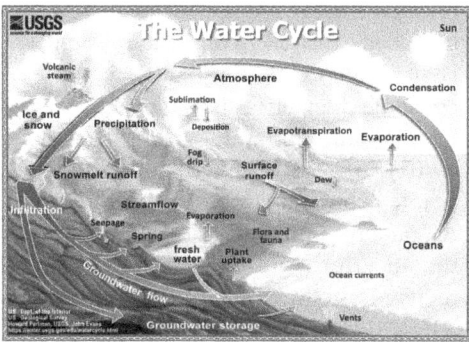

Figure 8.1 The hydro cycle, then and now (National Resources Board 1934 and US Geological Survey 2016).

arguing that schooling 'positively influenced' children's perceptions of the hydro cycle while the 'untutored ideas the children brought to school' (e.g. 'clouds are made by gods') negatively influenced their knowledge of water. Viewing the classic hydro cycle as the only 'accurate' understanding privileges Western epistemologies of water. Through these approaches, the hydro cycle has become a part of colonisation and globalisation processes, yet another means through which indigenous forms of environmental knowledge are marginalised.

Of course, not all educators adhere to this limited approach. Davies and Seimears (2008) suggest that 'unpacking' the multiple components of water (chemistry, biological function, societal uses, etc.) is necessary for teaching. Students, teachers, and groups can then pull ideas and issues together. Similarly, Eisen et al. (2009) use water as a case study for interdisciplinary teaching; their paper provides a guide for two water modules that combine the literature, art, philosophy, and science of water.

In many ways, pedagogy and educational practice are reflections of broader public discourse. Public participation around water-related discourses influences their knowledge of and interactions with water (Fosen 2012). For example, Moran (2008) discusses how people's engagement with composting toilets and greywater systems in their homes can motivate and catalyse policy around sustainability, while hidden, unacknowledged septic systems in the ground do not help improve awareness. At the same time, Stenekes et al. (2006) urge against supporting rhetoric that blames 'public ignorance' and 'cultural bias' for the failure of programs like water recycling, believing that this produces and reinforces a dichotomy between lay and expert opinions around water issues.

As Fosen (2012), Moran (2008), and Stenekes et al. (2006) demonstrate, public discourse does not stay stagnant over time, nor does academic discourse. The result is that many issues surrounding water that academics and research explore are not adequately conveyed by the classic hydro cycle. This is not to say that all hydrologic research is inevitably flawed, but rather that *the possibility exists*. Barnes (2001) reminds us that the dominant theories leading academic disciplines change over time. The classic hydro cycle was created in a time when epistemological theorising was dominant. However, Barnes (2001) suggests that hermeneutic theorising is gaining prominence in economic geography, focusing on interpretive and reflexive thinking and work. As underpinning theories and assumptions change in a discipline, so too should primary teaching and communication tools. The next section will review how thinking around water is changing in a number of disciplines, as scholars and practitioners place more emphasis on the dynamic interactions between human society and ecosystems.

Toward a more holistic approach: hydrosocial relations

While the classic hydro cycle has been a powerful tool for teaching and communication, its gaps are indeed many. Phrases like the 'hydro-illogical' cycle (Wilhite 2011) show that scholars now consider the classic model inadequate, and perhaps even harmful. Today's water knowledge is different than knowledge dominant as recently as the 2000s, let alone the 1930s. Beck's (1984) article, 'Topic of Public Interest: Water Quality', reflects on how post-war management impacted the collection and availability of empirical data and thus academics' ability to pay more attention to individual behaviour, pollution, and the like. His work serves as another reminder that our awareness has expanded since the hydro cycle was first conceptualised and created: The classic cycle cannot be expected to retain its power or accurately reflect what we know about water today.

This is not to say that we now know everything, or that a 'perfect' depiction of water movement could be created. Beven (2006, 609) highlights just how difficult hydrologic modelling is, acknowledging that in hydrology, the non-linearity, fluxes, and storages of water are hard to know entirely: 'The closure problem [boundary fluxes of mass, energy, and momentum in a watershed] is a scientific Holy Grail: worth searching for even if a general solution might ultimate prove impossible to find.'

Perhaps most obviously missing from the classic hydro cycle are humans: The vast majority of diagrams do not include a single individual or societal influence, appearing as animal- and human-free landscapes. This remains the case even though humans use, disrupt, redirect, and recycle water flows in a multitude of ways. Concepts like the 'precipitationshed', which refers to the upwind land and ocean from which rain in one area evaporated, recognise that actions in one area impact water availability, quality, and flows in other areas. Local actions have global impacts – global trends affect local issues – even as we continue to 'lack an adequate understanding of how the overall system works' (Vörösmarty et al. 2004, 513). In the introduction to the 'Geographies of Water', Fonstand (2013) points out that many of these human-induced changes are not new, but neither are they fully known. It is clear we need to better understand and respond to the connections between humanity, hydrologic flows, and ecosystems.

There is a 'complex web of interaction' featuring a great many feedbacks in human–environmental relations (Harden 2012). Castree (2002) sees a society–environment nexus (a networked series of connections) rather than dichotomy, shaped by a dialectical synthesis between humans and their environment. Other scholars engaged in socio-hydrology (Swyngedouw 2006), explore how 'water and society make and remake each other over space and time' through the hydrosocial cycle (Linton et al. 2014; Linton and Budds 2014), or posit the socio-cycle of water use and management (Turton 1999). These versions of the cycle include 'a flow not only of H_2O, but also one that is saturated with all manner of power relations' (Swyngedouw 2006, 15; see also Swyngedouw 2009). These schools of thought are more fully detailed in Schmidt's (2014) article 'Historicising the Hydrosocial Cycle'.

In opposition to Horton's vision of the hydro cycle as unbiased by scholars, Budds (2009) points to the wealth of literature arguing that physical assessments are not neutral and that studies are shaped by users' understandings. In 'Privatizing Water, Producing Scarcity: The Yorkshire Drought of 1995', Karen Bakker (2000) challenges conventional interpretations of the Drought, arguing that drought can be understood as the *production* of scarcity through the combination of meteorological modelling, demand forecasting, and corporate restructuring and regulation. The work of Budds, Bakker, and similar authors argues that the classic hydro cycle is the product of human framing rather than external, static, biophysical fact. Because of this, Budds (2009) argues that we need to consider both socio-political factors as well as geoclimatic ones in analysing waterscapes. We need to understand 'the ability and limits of freshwater ecosystems to respond to human-generated pressures' in the midst of 'altered hydrological regimes' (Naiman and Turner 2000, 958).

Natural scientists are beginning to do that reframing, considering quantitative hydrological science in a social context (see Sivapalan et al. 2012). According to Wesselink et al. (2016) socio-hydrology has the potential to bridge the gap between quantitative and qualitative measurements and may provide a baseline for hydrosocial analysis.

Graphical grappling: the UEA Hydro Cycle Working Group

Building on the emerging hydrosocial scholarship, a Hydro Cycle Working Group was convened by the Water Security Research Centre at the University of East Anglia. The interdisciplinary team included anthropologists, engineers, historians, and hydrologists from around the world with a mandate to identify gaps in the classic hydro cycle, determine which missing elements were most critical to communicate, and explore the creation of a new illustrative diagram more holistically capturing the way water moves.

The Group's discussions resulted in a list of some 50 geophysical aspects and more than 70 political, economic, and social considerations that would need to be included in a nuanced model of planet-wide hydrology (see Table 8.1 for a partial list of these issues). Research also demonstrates that various relatively simple alternative hydro cycles have been proposed by water activists and scholars. Consider, for example, the idea that 'water flows uphill to money' illustrated by Kate Ely in a modern, managed hydrologic system encompassing far more than geophysical processes (Ely n.d.; Figure 8.2).

The Hydro Cycle Working Group found two major concerns with these revised hydro cycles. One, the reworked images (of which Figure 8.3 is only one example) tend to focus on a single gap, illustrating only a few of the more than 120 elements identified by scholars as missing from the classic hydro cycle. Two, the alternative conceptions are far less prevalent. They have not been nearly as integrated in education, research, and policy and feel less intuitive – though this is likely a testament to how well the classic cycle is taught, not a reflection of the new images themselves. Neither of these concerns denies the value of hydro-cycle variations as teaching and demonstration tools, but they do suggest that the classic hydro cycle cannot simply be supplanted by one of these alternatives. The Group agreed that a major part of the classic hydro-cycle's power lay in its simplicity. A single, two-dimensional, static image will never manage to clearly but thoroughly communicate all of these important processes.

This logic led the Hydro Cycle Working Group to conclude that our original vision of a replacement image nuancing the hydrosocial cycle was unrealistic and, given the complexity of human–water interactions, perhaps undesirable. But during the course of those conversations, another idea was born: the creation of a toolkit that would enable communities, policy makers, researchers, or other populations to identify the elements of hydrology and human–water interactions most relevant to their water security, governance, management, access, and/or relations. Taking this idea, the compiled list of factors, and a library of alternative hydro-cycle images, the Hydro Cycle Working Group commissioned environmental artist Ruth Macdougall to begin working on the visual elements of the project. The resulting artwork is called the 'hydrosocial spiral'. While Ruth has developed a two-dimensional 'fixed' version of the hydrosocial spiral (Figure 8.3), the outcome is far more than a purely static image. Rather, the hydrosocial spiral is a participatory toolbox allowing for multi-dimensional representations of human–water processes and encouraging dialogue and reflection on water realities. In the next section, Ruth describes her iterative approach to creating and adding to the toolbox.

Table 8.1 Non-exhaustive list of elements to incorporate in hydrosocial models (generated by the Hydro Cycle Working Group)

Political, economic, cultural aspects	Geophysical aspects
Aesthetics	Advection
Agriculture	Aquifer storage and recovery
Biodiversity and environmental concerns	Climate change
Bioenergy	Condensation
Colours of water (blue, green, grey, etc.)	Deep percolation
Cloud seeding	Erosion and geological processes
Consumption patterns	Estuaries
Consumptive vs non-consumptive uses	Evaporation
Corporate vs national vs regional vs household vs individual uses of water	Glaciers
	Groundwater
Dams	Hydrofracking
Ecosystems goods and services	Hyporheic zone and flows
Efficiency	Infiltration
Gains, losses, and the paracommons	Interception
Fisheries and aquaculture	Macropores and flow
Industries	Ocean storage
Metaphysical and spiritual issues	Plant uptake
Outflow	Precipitation (rain, snow, sleet, hail, etc.)
Political borders	River discharge
Pumping	Rivers, lakes, streams, seas, oceans
Quality	Runoff
Pollution (acid rain, ocean plastics, etc.)	Saltwater intrusion
Recreation (water parks, swimming, etc.)	Snowmelt
Recycled water	Soil and rock drying, wetting, cracking, freezing, thawing, etc.
Manufactured reservoirs	
Rural vs urban use	Soil moisture
Securities (food, water resources, state, energy, community, economic, etc.)	Springs
	Storage
Transport	Sublimation
Waterways	Subsurface flows
Users and sectors	Terrestrial and aquatic ecosystems
Virtual water	Wetlands, coral reefs, etc.
Use in services	Thermal stripping
'Water flows uphill to money'	Transport
Water–energy–food nexus	Vapour, liquid, ice
Wastewater attitudes	Water table

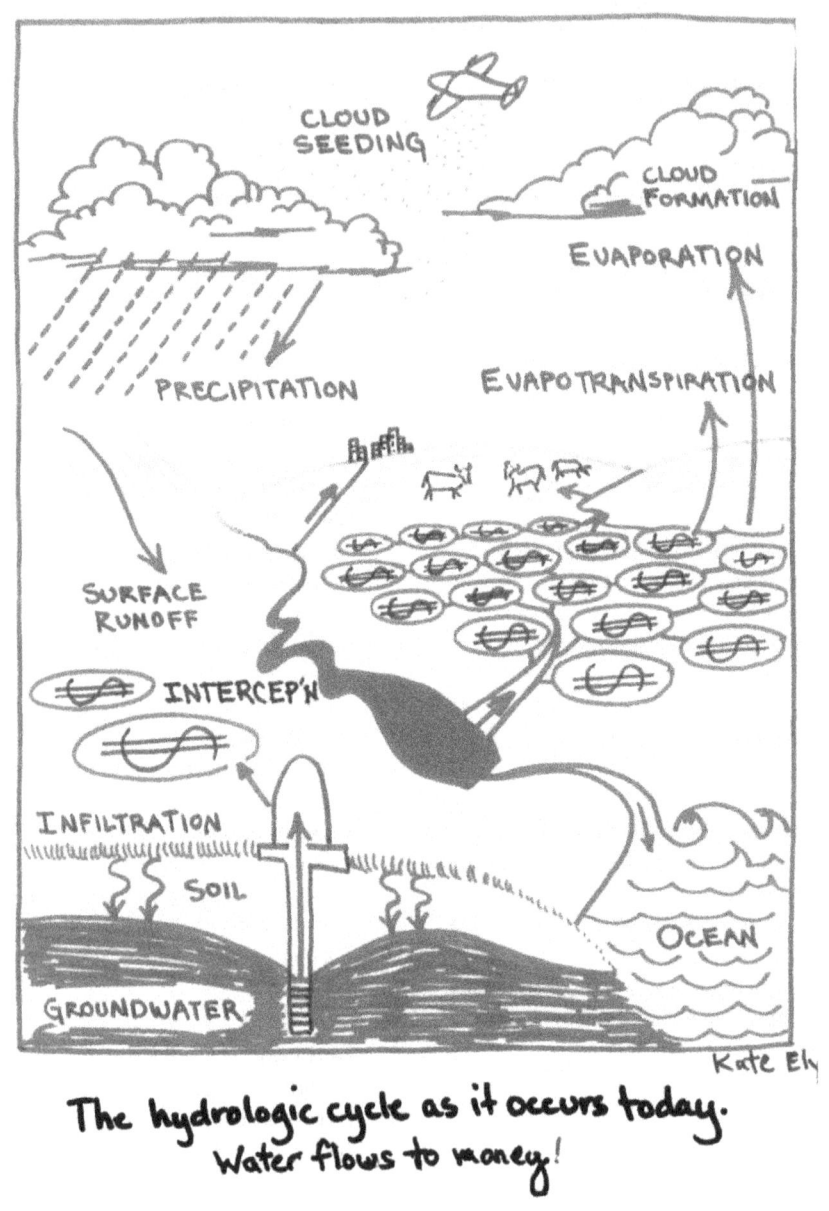

The hydrologic cycle as it occurs today.
Water flows to money!

Figure 8.2 Water flows uphill to money (Ely n.d., used with permission).

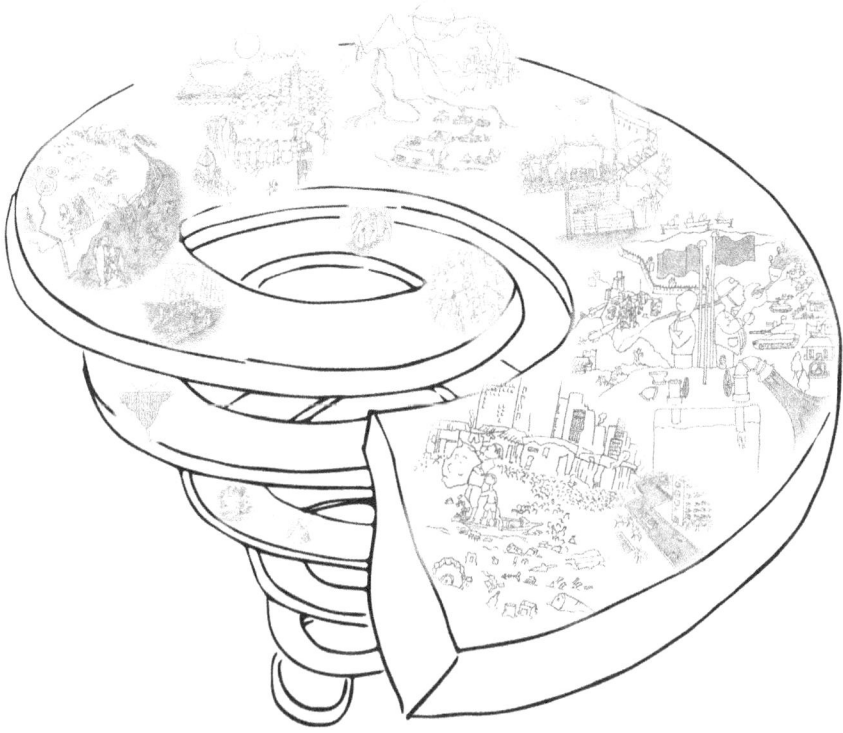

Figure 8.3 The hydrosocial spiral (created by Ruth Macdougall).

Developing the hydrosocial spiral: reflections and introductions from the artist

As with so many creative projects, the process that allows one concept to surface over many others may only be articulated upon reflection and with the help of more able wordsmiths than I. Through my research, I was introduced to Charles Minahen's (1992) *Vortex/t, the Poetics of Turbulence*, and at once recognised the key features of our spiral in his writing. In the appendix of his book, Minahen seeks to identify the features that make vortices, helices, spirals, and gyres distinct from each other while noting that these forms exist at all levels of the known universe, widely dispersed through the whole range of phenomena from macro to the macro cosmic, and inhering in both organic and inorganic systems and states.

The body we propose now is a 'spiral helix', a hybrid of the spiral and the helix, also known as a 'vegetal helix'. It exhibits circular movement with alternately increasing or decreasing circles evolving from a central point, towards new growth and around an invisible cone of energy. Importantly, the spiral helix shape allows us to move away from the closed loop of the hydro cycle and the use of arrows, which dominate almost all diagrammes associated with water. It is my opinion that these arrows disempower the viewer, exerting the limited meaning intended by the diagramme's authors, rather than allowing individual interpretations to be drawn by viewers living in vastly different environments. Instead, this structure allows multiple narratives to evolve.

The spiral helix is closely associated with power, an essential component of this new visualisation that not only references the kinetic power of water and whirl pools as depicted in the art of many cultures and creation myths, but also of the modern concept of water flowing upwards towards money and political power. As such, as the spiral winds round, time moves on, the environment changes. Population increases and so too does extraction. But there is one constant on the spiral: the outer edge represents those individuals and communities with political and economic power, and the inner edge represents those who suffer from a lack of political and economic power. It is proposed that a channel of blue water traveling along the spiral illustrates this power. In the earlier stage of the spiral (e.g. the bottom), the channel should travel almost centrally along the path of the spiral. Towards the top, as political power and inequity grow, the water channel dramatically veers to the outer rim.

In order to engage with the different scenarios taking place on the spiral helix, scale was the first obstacle to negotiate. As a result, a number of tableau were created in the form of movable discs, intended to be close-up vignettes of what may be occurring at various places on the spiral. Topics were chosen from the issues list identified by the Hydro Cycle Working Group (see Table 8.1). Unlike the classic cycle, these scenes place human activities firmly at the centre, depicting such themes as water engineering through aqueducts and dams, spiritual uses of water through purification rituals, and price tariffs through markets. Deforestation, climate change, agriculture, and recreation are also included (see Figure 8.4 for some example scenes) Numerous other vignettes could be included (e.g. desalination, cloud seeding, and hydro-diplomacy). The relative ease in adding topics through these movable discs is one way the hydrosocial spiral has proven useful for teaching. The next sections discuss some of our collective experiences in applying my work in water research and education.

Teaching the hydrosocial spiral: a participatory toolbox for co-learning

In Figure 8.3, the hydrosocial spiral is shown as a two-dimensional fixed image. Certainly these kind of static visualisations have their place in teaching, research, and policy. However, the greatest potential of the hydrosocial spiral lay in using it not as a diagram, but rather as a participatory tool. Using the basic structure of the spiral, groups – of students, community members, and policy makers – can be facilitated through a process of identifying the ways in

Figure 8.4 Example vignettes in the hydrosocial spiral. Top left: power inequalities mean
that the poor live and work amongst rubbish and polluted waters when cities
focus only on industry, commerce, and recreation. Top right: communities
protest the negative socioenvironmental impacts of dam infrastructure. Bottom
left: virtual water 'flows' between countries through the trade of food and
other water-intensive commodities. Bottom right: political conflicts arise over
transboundary water resources (created by Ruth Macdougall).

which water moves and changes in their particular setting. This visual tool
opens spaces for dialogue that purely verbal conversation may not; it also
provides a shared medium for individuals with different vantage points to
reflect upon. The authors suggest that the hydrosocial spiral can be of use as a
tool for research and policy, employed in focus groups to spark conversations.

When employing the hydrosocial spiral as a participatory tool, the suggested process involves introducing the history of the hydro cycle and

graphical modeling along with the basic logic of the spiral, breaking participants up into groups, and providing each group with a blank spiral along with cut out vignette and blank discs. Groups then develop their own version of the hydrosocial spiral, giving them the chance to highlight the processes they see as most impactful on issues of water availability and access. Depending on participants' local ecosystems, political situations, professions and livelihoods, and the like, vastly different discussions ensue.

Testing the participatory toolkit in classrooms at the University of East Anglia with postgraduate taught students, the biggest takeaway has been the incredible flexibility of the shape and discs. For the past four years, groups of water security students have been participating in three-hour seminars exploring these concepts and developing their own artistic representations of issues of interest. Year on year, the students have given serious thought to the spiral's construction and thoughtfully engaged with its strengths and limitations. The task invites the students to step outside their comfort zone, using their own hand-drawn images to investigate the cycle academically, artistically, and politically. Each time, the exercise brings conflicts of interest to the surface, with students pushing for varying foci dependent on their own background and preoccupations (conversations mirroring those of the original Working Group!).

During the construction of the hydrosocial spiral, it was of importance that the new visualisation be able to stand on its own and be easily understandable without further interpretation. Given that seminars have given student groups roughly 60 minutes to debate, decide, and create their artwork, the resulting visual responses are not always so easily legible, but they do represent a student-proclaimed helpful and reflective process.

Some students have instinctively seen the spiral in the same way its artist originally conceived of it, primarily as a representation of time and increasing technological intervention. Others have seen the spiral as a river leading toward a single human at the very bottom, and have drawn and placed various interruptions and interventions to that person's access to water along the spiral. Yet others have taken *two* spirals and woven them together to resemble the DNA double helix, using one to represent political processes and the other, ecological.

Still others have completely discarded the spiral. The challenge of developing an image or diagram that can speak to all environments is perhaps the greatest of all for building a global understanding of water. In dealing with this challenge, one team used a globe as the base visual. Another emphasised ecosystem variety by shifting from a global perspective to a comparative localised analysis. Their images were far simpler and more intuitive than the spiral, showing the way water is diversely used at the household level in Australia and Kenya with pictograms stylised as one might find in a children's book. Not only does their representation consider the ways in which two geographically different countries use and value water, it also brought into relief how political and economic differences impact everyday domestic life.

In feedback, this group emphasised that while they had not overtly used the spiral, their interpretation was inspired by it and the introduction of its development.

A particularly exciting divergent image created by students is shown in Figure 8.5. The group chose to focus on Dubai, a hydrological masterpiece created in the desert where skyscrapers and water parks abound. Their image moves away from the spiral, instead using the world's tallest skyscraper as the basis for an infographic. That symbol is far more relevant to the location – and speaks to the economic and political power held over (and by) water in that part of the world. Using the floors of the building to plot increasing and decreasing water issues over time, the format is an interesting one that leads to ideas of how water use in multiple cities could be explored using context-specific architecture. A 'hydrosocial city' could be created where water use in vastly different locations can be considered from across the street.

Even for those students who rejected the specific spiral, the general toolkit was a useful instrument. Groups made use of various vignettes and the idea of visual representation to inspire their own ideas and presentations. Overheard comments from students such as, 'It's very therapeutic to draw like this' and 'your brain is thinking in a different way', as well as solicited feedback praising the ease of critique via art suggests that visual media is a particularly effective tool in exploring complex ideas.

Using art allows the creativity of students, informants, and even researchers and policy makers to come forward. The ability to change, edit, and engage with the visuals gives participants a voice they may not otherwise have and allows them to identify what the original piece misses.

These lessons have also been applied in a Water School for primary school children in Southwest Morocco run by Dar Si Hmad, a local non-profit organisation harvesting fog to supply rural communities with potable water. In addition to learning the hydrologic science of the classic cycle, students use art to explore the different ways water moves in their villages. The approach allows them to understand the science behind fog-harvesting while valuing local knowledges and traditional approaches.

Taken together, these diverse classroom experiences indicate the strong potential of the hydrosocial spiral participatory toolkit as a method for engaging varied communities in discussions over water. The next section details a second application of the spiral: as a framework for analysing water issues.

Applying the hydrosocial spiral in research: a historic case study

As a tool to study hydrosocial interactions, the spiral may be more effective when applied to individual case studies rather than trying to address *everything* histori-cally and globally. In October 2015, a project on 'Reimagining Water Futures: Exploring Culture and the Communication of Water Stewardship Science' led

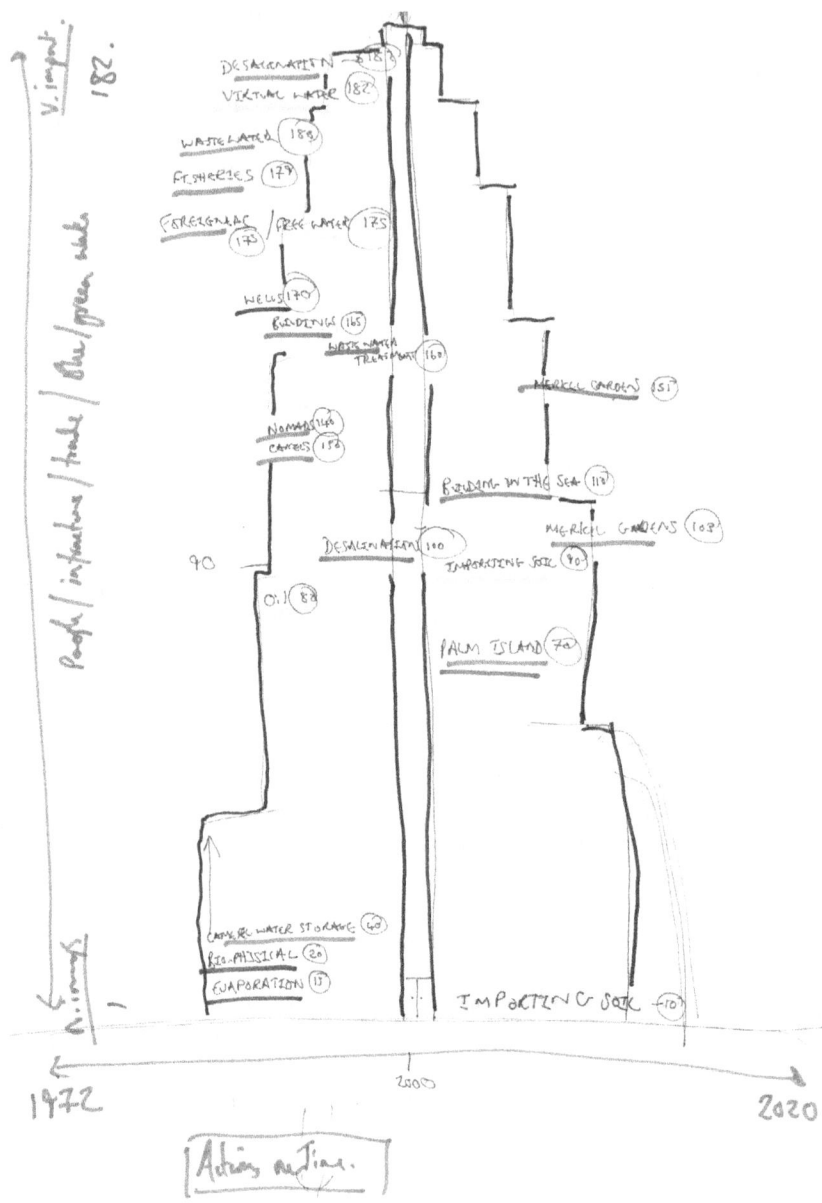

Figure 8.5 Sample artwork produced during a hydrosocial spiral seminar at UEA.

by Dr Naho Mirumachi at King's College London allowed Ruth to experiment with such an application. As part of the project, Ruth joined a workshop initiating networks between academics, science communicators, and cultural sectors. Mirumachi's approach reflected many of the aims of the new hydrosocial interactions scholarship discussions earlier, rejecting simplistic messages of 'water wars' or 'global water crisis' in favour of nuanced considerations of how local water problems are bound in issues of regional geopolitics, modern-day consumption, global food and energy trade, and power politics between (and within) the Global North and South. Working on the assumption that art has the power to affect ideational change and spur on those who have the catalytic ability to invoke that change (including businessmen and -women, consumers, politicians, and policy makers), Ruth introduced the hydrosocial spiral and discussed its development and applications with attendees.

As a consequence of the workshop, Ruth was able to advance her original conception and take a first step toward a three-dimensional incarnation of the spiral. Instead of discs, this 3D version utilises slides inserted onto the spiral. Just as groups are given blank discs to use with the 2D spiral, teams can be given empty slats for placement on a 3D model. These slats make visible the groundwater consequences of human extraction, pollution, politics, and the like, allowing the simultaneous consideration of both surface and subsurface activity over a period of time.

Figure 8.6 shows some of the first applications of this altered approach to the hydrosocial spiral, inspired by Battesti's (2012) 'The Power of Disappearance: Water in the Jerid Region of Tunisia'. Water usability and human interactions remain key elements to analysis. The slides explore the interactions and impacts of French colonisation, the Tunisian state, and local knowledge with oases in North Africa.

Figure 8.6 Advancing the hydrosocial spiral through movable slides (created by Ruth Macdougall).

Conclusions

Through hydrosocial relations approaches, water scholarship has progressed considerably since the classic hydro cycle was created in the 1930s. Unfortunately, the diagram has not evolved as fully as our thinking. This resulting gap provides a clear action step for further work. It is time to re-envision the hydro cycle. The hydrosocial spiral is an attempt to do just that in a dynamic, participatory way allowing for insight into the complexities of water and society.

From our experience, the spiral prompts more questions than it answers, underlining its efficacy as a catalyst for ideas rather than offering a defining visual solution. It is clear that continued and enhanced collaboration around water and hydro flows is necessary. This collaboration needs to be done in interdisciplinary settings and move beyond academia to include a wide range of stakeholders. Beyond the complexities of bridging academic disciplines, further study should also be done among people who view water in fundamentally different ways. Exploration should happen not only across the well-known *private versus public* and *human right versus economic resource* debates, but also between those who view water as a fundamental element and those who see it as a social construction. The human element of hydrosocial relations must be brought to bear even as the powerful non-human realities of nature are recognised. Multiple perspectives on water, some of which seem to be mutually exclusive, must be considered simultaneously and in conversation if we are to arrive at a more nuanced understanding of the hydro cycle and water itself.

As a hydrological diagram, the static hydrosocial spiral image is flawed. It is far more complex and non-intuitive than the classic hydro cycle and still fails to a number of key issues. But as a participatory toolkit, the approach has great potential. We continue to test that potential and ways to expand it. Water in our world is not static. Nor should the visualisations we use to understand and explore it be. A number of research and teaching tools on the hydrosocial spiral are available online at uea.ac.uk/water-security. The world's water is precious – but our diagram is not, so feel free to make use, undo it and challenge it, and be sure to let us know what you find.

References

Bakker, K.J. (2000) Privatizing Water, Producing Scarcity: The Yorkshire Drought of 1995. *Economic Geography*, 76(1), 4–27.

Barnes, T.J. (2001) Retheorizing Economic Geography: From the Quantitative Revolution to the 'Cultural Turn'. *Annals of the Association of American Geographers*, 91(3), 546–565.

Battesti, V. (2012) The Power of a Disappearance: Water in the Jerid Region of Tunisia. In Johnston, B.R., Hiwasaki, L., Klaver, I.J., Ramos-Castillo, A., and

Strang, V. (eds), *Water, Cultural Diversity, and Global Environmental Change: Emerging Trends, Sustainable Futures?* Paris: UNESCO/Springer SBM, pp. 77–96.

Beck, M.B. (1984) Topic of Public Interest: Water Quality. *Journal of the Royal Statistical Society Series A (General)*, 147(2), 293–305.

Beven, K. (2006). Searching for the Holy Grail of Scientific Hydrology: Asclosure. *Hydrology and Earth System Sciences*, 10, 609–618.

Brutsaert, W. (2012) *Hydrology: An Introduction.* Cambridge: Cambridge University Press.

Budds, J. (2009) Contested H_2O: Science, Policy and Politics in Water Resources Management in Chile. *Geoforum*, 40, 418–430.

Castree, N. (2002) False Antitheses: Marxism, Nature and Actor-Networks. *Antipode*, 34, 111–146.

Cockerill, K. (2010) Communicating How Water Works: Results from a Community Water Education Program. *Journal of Environmental Education*, 41(3), 151–164.

Davies, M.I and Seimears, C.M. (2008) Water: A Topic for All Sciences. *Science Activities: Classroom Projects and Curriculum Ideas*, 45(3), 27–36.

Eisen, A., Hall, A, Lee, T.S., and Zupko, J. (2009) Teaching Water: Connecting Across Disciplines and into Daily Life to Address Complex Societal Issues. *College Teaching*, 57(2), 99–104.

Ely, K. (n.d.) *Original Artwork ('The Hydrologic Cycle as It Occurs Today: Water Flows to Money!')*. Oregan: Confederated Tribes of the Umatilla Indian Reservation.

Fonstand, M.A. (2013) Geographies of Water. *Annals of the Association of American Geographers*, 103(2), 251–252.

Fosen, C. (2012) The Prism of Water: Environmental Rhetoric as Everyday Action. *Transformations: The Journal of Inclusive Scholarship and Pedagogy*, XXIII(2), 131–158.

Harden, C.P. (2012) Framing and Reframing Questions of Human–Environment Interactions. *Annals of the Association of American Geographers*, 102(4), 737–747.

Horton, R.E. (1931) The Field, Scope and Status of the Science of Hydrology. *Transactions, American Geophysical Union*, 12, 189–202.

Leopold, L.B. (1960) *Water: A Primer.* San Francisco: W.H. Freeman & Company.

Linton, J. (2008) Is the Hydrologic Cycle Sustainable? A Historical-Geographical Critique of a Modern Concept. *Annals of the Association of American Geographers*, 98 (3), 630–649.

Linton, J. (2010) *What Is Water? The History of Modern Abstraction.* Vancouver: UBC Press.

Linton, J. and Budds, J. (2014) The Hydrosocial Cycle: Defining and Mobilizing a Relational-Dialectical Approach to Water. *Geoforum*, 57, 170–180.

Linton, J., Budds, J., and McDonnell, R. (2014) The Hydrosocial Cycle: Mobilizing a Socio-Natural Concept. *Geoforum*, 57, 167–169.

McLaughlin, S.B., Wullschleger, S.D., Sun, G., and Nosal, M. (2007) Interactive Effects of Ozone and Climate on Water Use, Soil Moisture Content and Stream Flow in a Southern Appalachian Forest in the USA. *New Phytologist*, 174(1), 125–136.

Mienzer, O.E. (ed.) (1942) *Hydrology.* New York: Dover.

Minahen, Charles D. (1992) *Vortex/t: The Poetics of Turbulence.* Pennsylvania: Pennsylvania State University Press.

Moran, S. (2008) Under the Lawn: Engaging the Water Cycle. *Ethics, Place & Environment: A Journal of Philosophy & Geography*, 11(2), 129–145.

Naiman, R.J. and Turner, M.G. (2000) A Future Perspective on North America's Freshwater Ecosystems. *Ecological Applications*, 10(4), 958–970.

National Resources Board (1934) *A Report on National Planning and Public Works in Relation to Natural Resources Including Land Use and Water Resources with Findings and Recommendations*. Washington, DC: U.S. Government Printing Office.

Perrault, P. (1674) *On the Origin of Springs*. Reprint (1967). New York: Hafner.

Schmidt, J. (2014) Historicising the Hydrosocial Cycle. *Water Alternatives*, 7(1), 220–234.

Shepardson, D.P., Wee, B., Priddy, M., Schellenberger, L., and Harbor, J. (2009) Water Transformation and Storage in the Mountains and at the Coast: Midwest Students' Disconnected Conceptions of the Hydrologic Cycle. *International Journal of Science Education*, 31(11), 1447–1471.

Sivapalan, M., Savenije, H.H.G., and Blöschl, G. (2012) Socio-Hydrology: A New Science of People and Water. *Hydrological Processes*, 26(8), 1270–1276.

Stenekes, N., Colebatch, H.K., Waite, T.D., and Ashbolt, N.J. (2006) Risk and Governance in Water Recycling: Public Acceptance Revisited. *Science, Technology, & Human Values*, 31(2), 107–134.

Swyngedouw, E. (2006) Power, Water and Money: Exploring the Nexus. *United Nations Human Development Report Office Occasional Paper*. New York: United National Development Programme.

Swyngedouw, E. (2009) The Political Economy and Political Ecology of the Hydro-Social Cycle. *Journal of Contemporary Water Research & Education*, 142, 56–60.

Taiwo, A.A., Ra, H., Motswiri, M.J., and Masene, R. (1999) Perceptions of the Water Cycle Among Primary School Children in Botswana. *International Journal of Science Education*, 21(4), 413–429.

Turton, A.R. (1999) Water Scarcity and Social Adaptive Capacity: Towards an Understanding of the Social Dynamics of Water Demand Management in Developing Countries. *MEWREW Occasional Paper No. 9*. London: Water Issues Study Group, School of Oriental and African Studies.

United Nations Educational, Scientific and Cultural Organization. (1971) *Scientific Framework of World Water Balance*. Paris: UNESCO.

US Geological Survey. (2016, 15 December) The Water Cycle – USGS Water Science School. Available at: https://water.usgs.gov/edu/watercycle.html.

Vörösmarty, C., Lettenmaier, D., Leveque, C., Meybeck, M., Pahl-Wostl, C., Alcamo, J., Cosgrove, W., Grassl, H., Hoff, H., Kabat, P., Lansigan, F., Lawford, R., and Naiman, R. (2004) Humans Transforming the Global Water System. *EOS*, 85(48), 509–520.

Wesselink, A., Kooy, M., and Warner, J. (2016) Socio-Hydrology and Hydrosocial Analysis: Toward Dialogues Across Disciplines. *WIREs Water*, 4(2), 1196.

Wilhite, D.A. (2011) Breaking the Hydro-Illogical Cycle: Progress or Status Quo for Drought Management in the United States. *European Water*, 34, 5–18.

Chapter 9

Fluid-sound

Rob St John

This chapter outlines some possibilities for engaging sonically with freshwater aquatic systems. It focuses on the carrying streams of human and non-human sounds in and around water – described together here as *fluid-sound*. It offers approaches for engaging with what Hayden Lorimer (2005, 83) describes as 'our self-evidently more-than-human, more-than-textual, multisensual worlds'. Such 'non-representational' modes of engagement offer rich interdisciplinary potential for folding together techniques of geographical and artistic enquiry and production, as Phillip Vannini (2015, 3) puts it, 'a hybrid genre for a hybrid world'. Grounded in an attention to more-than-human[1] forces and processes, Derek McCormack (2015, 92) describes such exploratory methods as being about 'turning things around: defamiliarizing them; placing them in generative juxta-positionings that allow thinking to grasp a sense of liveliness of the worlds of things anew, however modestly'. Such approaches are less methodologies than styles or registers; experiential and experimental ways of being-in-the-world. This chapter is written through such registers, outlining a more-than-human approach for engaging with aquatic systems through sound, presenting three techniques for recording and collaboratively reimagining fluid-sounds.

We live in an age in which human influences have become all pervasive across global environments, shifting ecological and climatic processes and leaving traces at all scales: an era increasingly termed the 'Anthropocene'.[2] For Donna Haraway (2015, 160), the emerging epoch might be more appropriately termed the 'Chthuluscene', which 'entangles myriad temporalities and spatialities and myriad intra-active entities-in-assemblages – including the more-than-human, other-than-human, inhuman, and human-as-humus'. For Haraway (2015, 2016), ongoing socio-ecological concerns prompt us (as humans) to perceive, value and empathise with assemblages of life at all scales, a process of 'living together' in in that we may 'live and die well as mortal critters' (Haraway 2015, 160) As Helmreich (2015) notes, though, if we are to take seriously calls to attend to such more-than-human assemblages, we need to pay close and critical attention to the ways in which life is brought into perceptual being. What sound recording and production offers in this context, is an opportunity to engage with often unheard and unseen more-than-human worlds: sediment decomposition, pondweed photosynthesis, microinvertebrate activity, submerged flows, abstractions and bifurcations of water and pollutants.

Water flouts and transcends space, states, boundaries and categories; it seeps through complex and often leaky urban networks from our every pore through underground sewers to water-treatment plants, through ponds, canals, wetlands, bogs and rivers to pub cellars and swimming baths to our food, drinks and bodies. It can be alternatively clean and polluted; still and in spate; corralled and culverted in concrete yet always at risk of spilling over on to former flood plains. Water, particularly urban water, is a hybrid and dynamic medium, an often unseen and unnoticed support network for life, through which issues of capacity, resilience and adaptation are central in ecologically stressed worlds. Soundings[3] of water can be similarly mutable depending on the fluid state and space(s) of origin. For humans, listening underwater without the aid of technology can be a muffled and disorientating experience, lacking, as we do suitable organs to clearly receive underwater sound waves, as fish can through their lateral lines, or whales through specially adapted 'floating' ears (Helmreich, 2015). When frozen, water can take on a set of new acoustic properties, cracking and groaning as it physically shifts; variously dripping and popping as it melts[4]. Calm lakes and rivers can act as 'sound mirrors' for surrounding landscapes, reflecting and amplifying elements of their sonic traces.

The 'soundscape' concept is a central (and geographically inflected) means by which sound has been understood to emanate from, be shaped by, and shape the characteristics of different spaces and places. The term was developed and popularised by composer R. Murray Schafer in 1977. Schafer's framing of the soundscape as a means of understanding the 'tuning of the world' – namely its cultural and ecological components – was shaped by his concerns over the sonic impacts of noise pollution and environmental change. The centrality of such a perceiving subject in framing the soundscape (often through technological filters such as microphones) is clear in Emily Thompson's (2002, 1) contemporary definition of 'an auditory or aural landscape … simultaneously a physical environment and a way of perceiving that environment; it is both a world and a culture constructed to make sense of that world'.

In recent years, geographers have increasingly engaged with listening and sound-recording practices, leading to the nascent field of 'sonic geography' (Gallagher and Prior 2014; Gallagher 2015). Gallagher and Prior (2014, 2) suggest that 'audio recording produces distinctive forms of data and modes of engaging with spaces, places and environments which can function in different (and complementary) ways to more commonly used media such as written text, numbers and images', which can help researchers 'tell different kinds of stories to other media … particularly useful for highlighting hidden or marginal aspects of places and their inhabitants'. Attending to fluid-sound – both through listening and recording – may thus offer a valuable mode of multisensory research in multi-species landscapes (Tsing 2015), helping tease out 'small stories' of more-than-human worlds (H. Lorimer 2003), providing both audio material for 'creative geographies', which unsettle space-times (following Massey 1992; Hawkins 2013), and a document of the ongoing rhythms and becomings in/from a landscape.

The next section of this chapter narrates encounters with fluid-sound undertaken as artist-geographer. Drawing on the process-orientated philosophy of Tim Ingold and Gilles Deleuze, the listening practices outlined by Jean-Luc Nancy, and Henri Bergson's notion of duration, it is framed by reflective accounts of practices in two 'sonic geography' projects – Water of Life (Edinburgh, 2013) and Surface Tension (London, 2015). Three techniques for listening to, recording and remaking fluid-sound are explored:

1   Hydrophone;
2   Contact microphone;
3   Tape loop.

Recordings of each fluid-sound technique can be heard and downloaded at: www.robstjohn.co.uk/portfolio/fluidsound/.

Hydrophone

On the southern edge of Edinburgh is a small loch – a pond, really – in the middle of a housing estate; dug out of the footprint of a demolished tower block. On a sunny summer's day, the air above the water's surface flickers with life: a dragonfly resting on the skeleton of a partially submerged shopping trolley; a family of coot skittering through bankside irises, their leaves waving blue shopping bag flags in the breeze. The loch's shallow bed is thick with green pondweed, but seemingly still and calm. Lowering a pair of hydrophones into the tangle of oxygenating plants, a new aquatic lifeworld becomes audible. Percussive fizzing clicks ring out across the stereo field, the result of thousands of tiny air bubbles created through pondweed photosynthesis rising through the water column. A relentless 'busyness' invisible to the naked eye; a (quite literal) diurnal rhythm shaped by sunlight. Other sounds emerge from the backdrop of pops: the burring stridulations of underwater insects, the muted calls and cheeps of the coot family refracting through the surface tension and the regular upwelling of anonymous sounds in abstraction.

Hydrophones are underwater microphones that detect sub-surface vibrations and sound waves. They allow the recordist to hear underwater soundscapes that would otherwise be inaudible. Hydrophone technologies were developed through military use in the early twentieth century, largely in a marine setting, used as a complement for sonar and in so doing facilitating early recordings of whale song (Helmreich 2015). The resulting insights hydrophones offered into both the sounds of underwater life and the sonic characteristics of water have informed the work of artists such as John Cage, Max Neuhaus, David Dunn, Peter Cusack and Jana Winderen. In particular, Dunn's (2016, 28) environmental art practice figures hydrophone listening as means of accessing hidden more-than-human lifeworlds, suggesting that such practices can 'facilitate an increase in our collective environmental sensitivity and discovery of unknown natural and human made phenomena'.[5]

There is often a palpable sense of dislocation when underwater listening through hydrophones. Some sounds – like photosynthesis, the crackling and fizzing of icesheets, or the 'popping' noises made by marine shrimps in rock pools – are soon identifiable and strangely familiar, if rarely predictable. There are sounds that are harder to pin down: rumbles, drones, scratches; periodical intensities and silences. Some are artefacts of the hydrophone's movement on the waterbody's bed, often transmitting a sonic reflection of its materiality: the scree-slip sharpness of a gravel bed; the slow gloop of soft, stratified layers of silt. Others may be movements or sounds made by activity from underground trains or nearby industry, or the periodic abstraction or infilling of water to and from connected water networks. Perhaps the upwelling of trapped air from bottom sediments – a brief pulse of anaerobic microbial memory. Or, like a twitch on a fisherman's rod tip or float, the bow-waves of a passing shoal of fish.

Listening and recording with hydrophones has resonances with Bear and Eden's (2011) discussion of how recreational angling practices can create transformative encounters with animals, particularly fish. Drawing on Deleuze and Guattari's (1988) concept of the rhizome– characterised by a multiplicity of actors, processes and things – they write that 'becoming-fish' is 'not merely about the anglers' skilful mastery over a fish but about an affective contagion, involving an assemblage of fish, human and technology, each one already multiple. In other words, the angler and the fish 'enter into composition' with each other' (Bear and Eden 2011, 338). Critiquing Deleuze and Guattari's (1988) notion of 'becoming-animal' for its lack of curiosity about actual animals, Haraway (2008, 244) outlines an alternative notion of multispecies encounters: 'If we appreciate the foolishness of human exceptionalism, then we know that becoming is always becoming *with* – in a contact zone where the outcome, where who is in the world, is at stake'.

Hydrophones offer the potential for engaging with fluid worlds in ways akin to that Bear and Eden's angler; indeed the process of placing or throwing hydrophones beneath the surface is akin to 'fishing for sound'. However, following Haraway, we can also align it to emerging discussions around multispecies ethnography – 'how a multitude of organisms' livelihoods shape and are shaped by political, economic, and cultural forces' (Kirksey and Helmreich 2010, 545) – as a set of ongoing sonic becomings into which the listener/recordist is inherently folded. In the emerging field of multispecies enquiry, creative practices are foregrounded as a means of testing and troubling human/non-human boundaries and attending to the liveliness and agency of organisms at all scales (Kirksey 2014). There are resonances here with the styles and registers of non-representational methodologies, which similarly often incorporate creative practices in order to explore the multi-sensory and multi-model rhythms and affective[6] forces of the world (Vannini,

2015), and to the concerns of more-than-human geographies (Whatmore 2006), which acknowledge 'the lively agencies and hybrid ontologies of the nonhuman realm and ... construct accounts of human–nonhuman interaction which do not privilege human agency and consciousness' (J. Lorimer, 2009, 348).

Such an attentiveness towards the presence and potential agency of – often hidden water flows and lifeworlds – through sound has resonances with Gandy's (2004) notion of the 'urban metabolism' of water networks. Here, water – and its dynamic physical, chemical and symbolic elements and meanings – is foregrounded as an active agent in the production of space in urban areas. As a result, we might 'bring ashore' Steinberg and Peters' (2015, 248) oceanic 'wet ontology', to think with freshwater systems as 'a world of flows, connections, liquidities and becomings ... the reimagining and reenlivining of a world ever on the move'.

There are also echoes of debates in post-phenomenology (see review in Ash and Simpson 2016). In an essay 'Against Soundscape', Tim Ingold (2007) suggests that the soundscape concept frames environmental sound as something 'out there', which can be 'tuned into' by humans, when instead sound is inherently processional and co-produced by meshworks of human and non-human actors. For Ingold (2007, 11), listening to sound is instead the process of 'immersion in, and commingling with, the world in which we find ourselves'. Helmreich (2010) extends this argument, stating that the soundscape concept 'emerges from a mix of contemplative aesthetics and technologies of objectification and subjectification. The soundscape is shadowed by an acoustemology of space as given and listener as both apart from the world and immersed in it'. Acoustemology is the term coined by Steven Feld (2000, 184) to describe human engagement with sound through 'a union of acoustics and epistomology ... the primacy of sound as a modality of knowing and being in the world'.

In order to overcome this phenomenological conceptualisation of a soundscape 'out there', which is variously heard, felt and described by human perception, Helmreich proposes a transductive concept of listening underwater. For Helmreich (2007, 622), a transductive approach questions the 'cognitive, affective, and social effects of transducing – that is, converting, transmuting – sound from the medium of water into that of air'. Listening underwater is thus the simultaneous 'mixed' sensing of medium and matter, mediated through listening technologies such as hydrophones and contact microphones.[7] This approach shares a post-phenomenological perspective with Ingold (2008), in which human perception of the world is continually 'coming into being' through a constant binding and unbinding of its surfaces, substances, airs and atmospheres. For Ingold (2008), such 'geographies of mixture' are characterised by ongoing processes of amalgamation, distillation, coagulation and dispersal, in a manner akin to Mol and Law's (1994) notion of 'fluidspace', in which there are no well-defined objects or entities: rather a flow of substances that accumulate in temporary, ephemeral forms. Transduction, then, may offer a means of understanding listening to fluid-sound in a way that unsettles fixed structures and boundaries.

The experience of engaging with water flows through transductive and interdisciplinary sound practices might be framed in terms of Deleuze's concept of the 'encounter'. For Deleuze (1968 [1994], 139), the encounter is 'something in the world [that] forces us to think', an affective force that causes a break or rupture in our habitual ways of interacting with the world, and forces us to undergo reflection or reconfiguration of these interactions. Jane Bennett (2012, 232) expresses a similar sentiment, suggesting that encounters with the world through creative processes (in this case, poetry) can 'help us feel more of the liveliness hidden in such things and reveal more of the threads of connection binding our fate to theirs'. Engaging with fluid-sound with hydrophones offers the potential to open up new spaces of encounter, which might help us perceive, comprehend and think with the different assemblages of more-than-human life with which we are intertwined (Dixon et al. 2012). Listening with hydrophones foregrounds the notion of the 'soundscape' as a collaborative, ongoing process between its source, the (technological) transductor, and (human) receiver; sounds that can alter along micro-scales afforded by minor position shifts, revealing invisible transects of diversity. Listening to fluid-sounds offers the potential to attend to the multiple flows, processes and soundings of worlds woven through by water, and typified by temporary and sometimes uncertain human and non-human forces: splits and joins, currents and eddies, suspensions and dissolutions, becomings and disappearances.

Contact microphone

The trail begins in the city archives; following the trail of the first water systems through Edinburgh; water sprung from the Pentland hills, and run through wooden pipes to the growing city. In the late 1600s, a series of springs were routed to a single wellhouse, and each of the spring wellheads marked by a lead sculpture of an animal: a fox, a swan, a lapwing and a hare (rumours of an additional owl remain inconclusive). The stone wellhouse, a small chapel-like structure with a triangular roof, has been out of use since the 1940s, a relic of a pre-industrialised era of water management. Standing in a small playing field bordering the city bypass, the structure has a resurgent vegetative upwelling of its own, as nettles and brambles tangle towards the sky, entwined with shaky security barriers; a circle of unruly life where the park's mower can't reach. The spectre of the sculptures (taken to the city museum in the 1960s) lends a sense of animism as much as the water does a sense of animation. Whatever, for now, the wellhouse is off limits; quiet in the midsummer bloom. Teetering on the lip of an unstable metal fence, I reach its heavy black metal door; which barely sounds when tapped with a fist. A pair of small contact microphones, each roughly the size of a two-pence piece, are jammed into the gaps between the metal door and its stone frame. Transmitting through trailing black wires in the undergrowth to headphones laid out on the grass, the building's subtle rumbles and resonances are suddenly audible: the unmistakable ebbing flow of water; less a roar than an yawning/becoming of water springing forth. The ruin reframed by flow: as ongoing, as emergent.

Contact microphones (or piezos) sense audio vibrations through contact with solid objects: transductors of sound waves and physical movements mediated through material. They can be variously clipped (with woodworking clamps), stuck (with electrical tape) or wedged (into cracks in rock or bricks) into material. Contact microphones thus offer the practitioner the potential to access hidden rhythms of the world, means of abstracting invisible water flows across all manner of surfaces – a 'lost' urban river running under a manhole cover, the rattle of a long fence wire caught by flood water, the bell tones of a submerged dock cable chain, the braided trails of water through a ruined building.

Jean-Luc Nancy's (2007) distinction between the processes of 'listening' and 'hearing' is a useful means of conceptualising the use of contact microphones. For Nancy, to hear (*entendre*) is to understand and contextualise a broadly recognisable sound: he suggests 'a siren, a bird, or a drum' (ibid., 6). On the other hand, to listen (*écouter*) is 'to be straining toward a possible meaning, and consequently one that is not immediately accessible' (ibid., 7), whether an unfamiliar sound, piece of music or sound art, or ideas in speech. The listening process, for Nancy, is 'one where sound and sense mix together and resonate in each other, or through each other' and where 'to be listening is always to be on the edge of meaning . . . a resonant meaning, a meaning whose *sense* is supposed to be found in resonance, and only in resonance' (ibid.).

Listening and recording with contact microphones can detect otherwise inaccessible rhythms and resonances through material; in so doing, opening up spaces of possibility, encounter and meaning. Sounds abstracted from their source can generate a (sometimes tense) range of affective intensities somewhere between curiosity, imagination and frustration. As John Cage (1952/1967) might have put it, the listener is prompted to hear such sounds 'in themselves', stripped of referentiality. But, following Nancy, such strained listening to unheard resonances of the world may create fertile proving grounds for extending human perception of our entanglements with more-than-human worlds through fluid-sound.

Contact microphones are frequently used in experimental music and sound art practices as a means of teasing otherwise-inaccessible sounds from everyday objects and instruments (Gottschalk 2016). They are cheap and straightforward to build with a basic understanding of DIY electronics. Contact microphones have a unique capacity to access and record emergent sounds, rhythms and resonances in the landscape. Seemingly inert materials can be heard as animated, lively and fluid through contact microphones, as they variously transmit, filter and refract audio vibrations. Water as a weathering, sculpting, shaping, percolating, seeping, seeding agent of change is foregrounded here by sound. There is vast potential, as a result, for practitioners seeking to engage with post-human notions of the agency and fluidity of 'inert' non-human materials and objects such as rocks (e.g. Bennett 2009; Cohen 2015; Dixon et al. 2012; Yusoff 2014), through contact microphones. Contact microphones, in short, can help reveal the seemingly inert as inherently lively.

Tape loop

A bucket of Lea river water is carried back on the train from London to Lancashire; fluorescent specks of duckweed caught around the rim like abandoned archipelagos, deltas of minor spills seeping along the carriage floor. Split across a series of darkroom trays, the water – collected from a stretch of the lower river lit by eutrophic, oily slicks (organic traces of the city's excesses) – becomes a developing bath for sounds recorded in and around the Lea. Recordings of the river's soundings are dubbed on to 1/4' tape using a reel to reel recorder, and loop short lengths of tape: top to tail; source to mouth. Each loop roughly the length of outstretched arms, carrying a few seconds of abstracted sound embedded in a magnetic layer. Shimmering like elvers, the loops are left for varying lengths of time in the developing baths – a day, a week, a month – and then dried and replayed. The action of the water, its organic life and dissolved pollutants gradually alter and erode the surface of the tape, leaving thin threads of discontinuity: braids and knickpoints. When replayed, some of the 'developed' tapes transmit more noise than signal; the recording only haunting the edges of white noise, generated here (perhaps counter-intuitively) by still, rather than turbulent waters. A small number are transformed in a different way. A loop revolves: the white tape marker of the join – like a depth marker on a river bank – travelling out from the spool, around a heavy glass bottle of water, and back again. Each orbit shifts subtly; new rests and reconfigurations of a fracturing melody and pealing rhythm, until finally, a submerged sort of silence.

There are numerous ways of bringing recordings of fluid-sound together in forms that might be used in multi-media academic/artistic publications, presentations, installations and releases, drawing from thick (and often entwined) histories of sound art and music (Connor 2010; Gough 2016). One such technique is the tape loop. Recordings made with hydrophones, contact mics and other sources can be dubbed onto magnetic tape, which is then cut into sections and spliced into loops. Tape loops can be physically altered by their durational immersion in aquatic environments: a technique that might be understood as an experimental collaboration with more-than-human forces and processes to create sound pieces for which the recordist has only set the starting points for emergence.

Such practices owe a debt to New York artist and composer William Basinski. In 2001, Basinski began digitising a series of tape loops[8] of synthe-siser music he had made in the early 1980s. Replaying the loops on a reel-to-reel recorder, Basinski (2001) found that intervening years had caused the tapes' magnetic strips to decay and warp:

I soon realized that the tape loop itself was disintegrating: as it played round and round, the iron oxide particles were gradually turning to dust and dropping into the tape machine, leaving bare plastic spots on the tape, and silence in these corresponding sections of the new recording ... Yet the essence and memory of the life and death of this music had been saved: recorded to a new media, remembered.

His resulting work, *Disintegration Loops*, consists of six separate loops repeating and decaying for anything up to an hour, each altering in substantially different ways.[9] Basinski draws a direct link between the fallible materiality of the magnetic tape layer and the slow decay of human memory and bodies (Gough 2016). The magnetic tape has a lively 'body', which is slowly reduced to flaking dust by the action of playback, an ever-eroding sense of granularity and atomism revolving in creative-destructive tension.

Disintegration Loops signposts the potential of using tape loops as a means of engaging sonically – and non-representationally – with aquatic lifeworlds. In the vignette above, recordings of the River Lea in London – made using contact and hydrophone microphones – were dubbed onto tape alongside sparse cello and piano instrumentation responding to their fluctuations. The Lea is one of the most polluted and modified rivers in Britain. The combined (if difficult to quantify) action of microbial decomposition, the physical effects of submergence in water, and the effects of suspended and dissolved inorganic pollutants etch and decay the 'body' of the magnetic tape, adding new layers of process and alteration to a discrete snippet of space and time encoded in sound. Working with environmentally 'developed' tape loops offers possibilities for recorded fluid-sounds to be shaped and altered in an emergent process of creative destruction tied to the biological and chemical processes of a particular place or space: a polluted water bath, an anaerobic bog, a discarded cup of morning coffee.

As ongoing, emergent moments of space-time caught between accumulation and disintegration, repetition and change, tape loops are a generative means of engaging with water flows, processes and lifeworlds through sound. For Connor (2010, 4), 'loops are parentheses, procrastinations, pockets of time and space which are held apart from the general conditions of propagation and passing away ... a loop saturates space, filling it up from the inside out'. In the constant state of 'feedback' in tape loops, Gough (2016, 95) writes that, 'individuals and objects, or individuals and events, mobilize each other; they are both agents. Both the discrete instances and continuous processes are necessary for history to "run"'. Loops, then, can trouble time and space and muddle intention and outcome; oscillating in a constant state of tension and reciprocal sonic change and becoming.

Here, Henri Bergson's (1946) notion of duration (or *la durée*) is useful for conceptualising the use of disintegrating tape loops. For Bergson, time should be conceived as a flow or continuum, in which the past, present and future permeate each other, producing beings and events in an ongoing processional 'becoming'. In short, time is always mobile and incomplete; a multiplicity of ephemeral moments of different rhythms, vibrations, tensions, dilations and contractions. The revolving tape loop is a procession of sonic moments, each shaped by historical conditions and choices (the environment, the recorder, the recordist, the tape), but never quite the same; an ongoing echo of worldly spaces, processes and conditions; a durational topology of chance.

A pre-history to such approaches can be found in auto-destructive art. Writing in 1960, Gustav Metzger, the chief architect of the movement states, 'Auto-destructive art demonstrates man's power to accelerate disintegrative processes of nature and to order them ... Auto-destructive art is the transformation of technology into public art' (cited in Stiles and Selz 2012, 470). Through the use of acid paints, smashing glass and entropic sculptures, 1960s artists like Metzger and Jean Tinguely pointed to how destructive artistic processes could (perhaps counter-intuitively, initially) inform a new mode of creative practice and production, closely engaged with socio-political and ecological concerns. More recently, the artist-academic Daro Montag (2001) has created a series of 'bioglyphs' using photographic negatives and paper, artefacts traced by the activities of micro-organic life, created by burying films in earth or placing organic matter such as decaying fruit upon their surface. Similarly, the artist and musician Richard Skelton created a series of recordings (e.g. *Landings*, 2009) based around a series of instruments buried on the West Pennine Moors, then exhumed and played.

The concept of emergence is central to such auto-destructive (or, in many senses auto-*creative*) work, particularly that of Montag and Skelton. Emergence is an organising principle in which larger or more complicated phenomena arise through interactions among smaller or simpler phenomenon (Johnson 2001). The emergent phenomena cannot be reduced to the properties or characteristics of their constituent parts. Emergence is a concept that takes on different registers and meanings in different disciplinary realms, and is central to Henri Bergson's (1946) vitalist philosophy.[10] Thinking with emergence can help us see the way that other things – pondweed, water or sediment, say – have powerful agency in shaping worlds, both vast and tiny, and that our lives as humans are forever caught up in these entangled webs of coexistence. Allowing for emergence in research and practice creates space to be challenged, surprised or disappointed. Starting points for enquiry and process are set – durational recordings taken in a space, place, transect or drift; technological means for engaging with the rhythms of the world – and then potentially unpredictable (in)organic patterns and processes are encouraged. 'Environmental' production techniques, of which ecologically altered tape loops are only one,[11] offer the potential for the out-comes of such a 'sonic geography' to be determined by worldly processes such as disintegration and decay; perhaps aptly in an age of socio-ecological uncertainty, flux and environmental change.

Practically, such tape-loop approaches require modest amounts of equip-ment and expenditure. Their primary investment for the artist-geographer is one of time: to make, edit and dub recordings onto tape; to conceptualise and carry out a fluid alteration or degradation process (if deemed appropriate); and to play back the resulting sound loops over sufficient duration that any shifts in their character, intensity and affectual resonances become apparent. This is, evidently, slow work with no guarantee of emergent outcomes. But, such slowness and uncertainty of outcome can lend a productive tension to the research process, and particularly to any (re)presentation of the work in installations, presentations and releases. As a disintegrating tape loop shifts over the duration of its playback it initiates both the practitioner and listener (*sensu* Nancy) into an unfolding sense of attentive, or 'strained', listening. In other words, through the durational process of replaying a dynamic loop of sound, meaning is constantly accumulating, fragmenting and disintegrating through a series of resonances and referrals. The listener becomes inherently folded into the fluid dynamics of recorded spaces and human and non-human traces 'becoming' (*sensu* Bergson) through sonic forms in unstable processes mediated by the collective actions of the artist-researcher and the mediating technologies.

Looping and (re)sounding

Drawing on experiences and experiments as an artist-geographer, this chapter has traced a set of techniques for interdisciplinary engagements with aquatic environments using sound. It highlights the potential of drawing from thick lineages of creative practice to inform and extend contemporary geographical and socio-ecological debates, events and (non)representations. As McCormack (2015, 100) writes, this involves

> taking a familiar technique from one context [in this case sound art] and showing how it can do a qualitatively different kind of work in another, and in a way that remakes that technique, or inventively inflects it, or transforms it such that both it and world in which it is situated are rendered strange.

A common thread to all of the processes described here is the role of more-than-human assemblages of humans, non-humans, materials and technologies enrolled in the co-production (and potential disintegration) of fluid-sound. In many ways, this approach represents a break from many established modes of sound recording, concerned with clarity, fidelity and the isolation of individual sounds.[12] In effect, the techniques here form part of ongoing conversations about how socio-ecological and geographical researchers may become 'creative' or 'experimental' in their practice (e.g. Hawkins 2013), and particularly in how their work may attend to the non-representational rhythms, affects and intensities of the world (H. Lorimer 2015). The spirit of this chapter, then, is to encourage ongoing experiments to expand to possibilities of creative research and practice through fluid-sound.

Acknowledgement

Thanks to Tom Western, Jared Margulies, Emma Cardwell, Deborah Dixon, Minty Donald and three anonymous reviewers for invaluable comments on drafts of this chapter.

Notes

1 Defined by Sarah Whatmore (2006, 604) as 'modes of enquiry [that] neither presume that socio-material change is an exclusively human achievement nor exclude the "human" from the stuff of fabrication [and] attend closely to the rich array of the senses, dispositions, capabilities and potentialities of all manner of social objects and forces assembled through, and involved in, the co-fabrication of socio-material worlds'.

2 For original Anthropocene terminology, see Crutzen and Stoermer (2000); for a post-colonial reading of how the concept unsettles imaginaries of the 'human', see Chakrabarty (2009); and for a folding of process-orientated cultural theory and conservation biology in the Anthropocene, see J. Lorimer (2015).

3 Helmreich (2015: xi) reframes the oceanographic term *sounding* as one that is useful 'for investigating things not yet known, things whose limits are not clear or whose boundaries may be obscured – perhaps by the sounding apparatus itself'.

4 The sound of melting ice-caps and calving glaciers might be figured as a 'soundmark' (analogous to a distinctive terrestrial 'landmark' (in Schafer's (1977) terms) of ongoing climate change; explored in sonic works including Chris Watson's *Weather Report* (2004) and Katie Paterson's *Vatnajökull* (2007).

5 Dunn's (2016) work on using contact microphones to sense (and, ultimately manage) the destructive agency of pinon bark beetles in California is a notable example of such art-ecologies in practice.

6 H. Lorimer (2008, 552) describes affects as 'properties, competencies, modalities, energies, attunements, arrangements and intensities of differing texture, temporality, velocity and spatiality, that act on bodies, are produced through bodies and transmitted by bodies'.

7 See Sterne (2003) for a discussion in the role of technology in mediating listening practices and cultures.

8 Repeating loops of audio tape; used by mid-twentieth-century music concrète and minimalist composers such as Pierre Schaeffer, Steve Reich, John Cage and Karlheinz Stockhausen.

9 See https://williambasinski.bandcamp.com/album/the-disintegration-loops.

10 See Ash and Simpson (2016) for a review of the centrality of emergence in post-phenomonological approaches; Kirksey (2015) for a review of the concept of emergence through non-equilibrium ecologies, dynamic conservation approaches and process-based environmental art; Bateson (1972) for a cybernetics-inflected exploration of emergence in social and ecological systems.

11 Digital audio loops can be produced using software such as Max MSP. Similarly, sonification is a popular emerging means of making worldly rhythms, patterns and process audible through the 'sonification' of datasets (e.g. Palmer and Jones 2014), but there remain significant questions over the tensions between the aesthetic qualities, affective responses and actual representations of the world that they generate.

12 See the variety of contemporary sound recording perspectives and practices offered in Lane and Carlyle (2013).

References

Ash, J. and Simpson, P. (2016) Geography and Post-Phenomenology. *Progress in Human Geography*, 40(1), 48–66.

Basinski, W. (2001) Disintegration Loops. http://gewissheit-vision.de/basinski_en. php. Accessed 17 March 2017.

Bateson, G. (1972) *Steps to an Ecology of Mind: Collected Essays in Anthropology, Psychiatry, Evolution, and Epistemology*. Chicago: University of Chicago Press.

Bear, C. and Eden, S. (2011) Thinking like a Fish? Engaging with Nonhuman Difference through Recreational Angling. *Environment and Planning D: Society and Space*, 29(2), 336–352.

Bennett, J. (2009) *Vibrant Matter: A Political Ecology of Things*. Durham, NC: Duke University Press.

Bennett, J. (2012) Systems and Things: A Response to Graham Harman and Timothy Morton. *New Literary History*, 43(2), 225–233.

Bergson, H. (1946) *The Creative Mind: An Introduction to Metaphysics*, trans. M.L. Andison. New York: Citadel.

Cage, J. (1952/1967) Juilliard Lecture. In *A Year from Monday: New Lectures and Writings by John Cage*. Connecticut: Wesleyan University Press, pp. 95–112.

Chakrabarty, D. (2009) The Climate of History: Four Theses. *Critical Inquiry*, 35(2), 197–222.

Cohen, J.J. (2015) *Stone: An Ecology of the Inhuman*. Minneapolis: University of Minnesota Press.

Connor, S. (2010) A Philosophy of Fidgets. Liverpool Biennial Touched Talks, 17 February 2010, http://stevenconnor.com/fidgets/fidgets.pdf. Accessed 17 March 2017.

Crutzen, P. and Stoermer E. (2000) The 'Anthropocene'. *Global Change Newsletter*, 41, 17–18.

Deleuze, G. (1968/1994) *Difference and Repetition*, trans. P. Patton. New York: Columbia University Press.

Deleuze, G. and Guattari, F. (1988) *A Thousand Plateaus*. London: Athlone.

Dixon, D.P., Hawkins, H. and Straughan, E. (2012) Of Human Birds and Living Rocks: Remaking Aesthetics for Post-Human Worlds. *Dialogues in Human Geography*, 2(3), 249–270.

Dunn, D. (2016) A Philosophical Report from Work-In-Progress. In F. Bianchi and V. J. Manzo (eds), *Environmental Sound Artists: In Their Own Words*. Oxford: Oxford University Press, pp. 27–33.

Feld, S. (2000) Sound Worlds. In P. Kruth and H. Stobart (eds), *Sound*. Cambridge: Cambridge University Press, pp. 173–198.

Gallagher, M. (2015) Field Recording and the Sounding of Spaces. *Environment and Planning D: Society and Space*, 33(3), 560–576.

Gallagher, M. and Prior, J. (2014) Sonic Geographies: Exploring Phonographic Methods. *Progress in Human Geography*, 38(2), 267–284.

Gandy, M. (2004) Rethinking Urban Metabolism: Water, Space and the Modern City. *City*, 8(3), 363–379.

Gottschalk, J. (2016) *Experimental Music Since 1970*. London: Bloomsbury.

Gough, K.M. (2016) The Art of the Loop: Analogy, Aurality, History, Performance. *TDR/The Drama Review*, 60(1), 93–115.

Haraway, D.J. (2008) *When Species Meet*, Minneapolis: University of Minnesota Press.

Haraway, D.J. (2015) Anthropocene, Capitalocene, Plantationocene, Chthulucene: Making Kin. *Environmental Humanities*, 6(1), 159–165.

Haraway, D.J. (2016) *Staying with the Trouble: Making Kin in the Chthulucene*. Durham: Duke University Press.

Hawkins, H. (2013) *For Creative Geographies: Geography, Visual Arts and the Making of Worlds*. London: Routledge.

Helmreich, S. (2007) An Anthropologist Underwater: Immersive Soundscapes, Submarine Cyborgs, and Transductive Ethnography. *American Ethnologist*, 34 (4), 621–641.

Helmreich, S. (2010) Listening Against Soundscapes. *Anthropology News*, 51(9), 10–10.

Helmreich, S. (2015) *Sounding the Limits of Life: Essays in the Anthropology of Biology and Beyond*. Princeton: Princeton University Press.

Ingold, T. (2007) Against Soundscape. In E. Carlyle (ed.), *Autumn Leaves: Sound and the Environment in Artistic Practice*, Paris: Double Entendre, pp. 10–13.

Ingold, T. (2008) Bindings against Boundaries: Entanglements of Life in an Open World, *Environment and Planning A*, 40(8), 1796–1810.

Johnson, S. (2001) *Emergence*, London: Penguin.

Kirksey, E. (ed.) (2014) *The Multispecies Salon*. Durham: Duke University Press.

Kirksey, E. (2015) *Emergent Ecologies*. Durham: Duke University Press.

Kirksey, S. and Helmreich, S. (2010) The Emergence of Multispecies Ethnography. *Cultural Anthropology*, 25(4), 545–576.

Lane, C. and Carlyle, A. (2013) *In the Field: The Art of Field Recording*. Axminster: Uniformbooks.

Lorimer, H. (2003) Telling Small Stories: Spaces of Knowledge and the Practice of Geography. *Transactions of the Institute of British Geographers*, 28(2), 197–217.

Lorimer, H. (2005) Cultural Geography: The Busyness of Being More-Than-Representational. *Progress in Human Geography*, 29(1), 83–94.

Lorimer, H., (2008) Cultural Geography: Non-Representational Conditions and Concerns. *Progress in Human Geography*, 32(4), 551–559.

Lorimer, H. (2015) Afterword: Non-Representational Theory and Me Too. In P. Vannini (ed.), *Non-Representational Methodologies: Re-Envisioning Research*. Abingdon: Routledge, pp. 177–189.

Lorimer, J. (2009) Posthumanism/Posthumanistic Geographies. In R. Kitchin and N. Thrift (eds), *International Encyclopedia of Human Geography, Vol. 8*. Oxford: Elsevier, pp. 344–354.

Lorimer, J. (2015) *Wildlife in the Anthropocene*. Minneapolis: University of Minnesota Press.

McCormack, D.P. (2015) Devices for Doing Atmospheric Things. In P. Vannini (ed.), *Non-Representational Methodologies: Re-envisioning Research*. Abingdon: Routledge, pp. 89–111.

Massey, D. (1992) Politics and Space/Time. *New Left Review*, 196, 65.

Mol, A. and Law, J. (1994) Regions, Networks and Fluids: Anaemia and Social Topology. *Social Studies of Science*, 24(4), 641–671.

Montag, D. (2001) *Bioglyphs*. Totnes, UK: Festerman.

Nancy, J.L. (2007) *Listening*, trans C. Mandell. New York: Fordham University Press.

Palmer, M. and Jones, O. (2014) On Breathing and Geography: Explorations of Data Sonifications of Timespace Processes with Illustrating Examples from a Tidally Dynamic Landscape (Severn Estuary, UK). *Environment and Planning A*, 46(1), 222–240.

Paterson, K. (2007) *Vatnajökull* [Sound Installation]. London: Slade School of Fine Art.

Schafer, R.M. (1977) *The Tuning of the World*. New York: Alfred A. Knopf.

Skelton, R. (2009) *Landings* [CD]. Sustain-Release.

Steinberg, P. and Peters, K. (2015) Wet Ontologies, Fluid Spaces: Giving Depth to Volume Through Oceanic Thinking. *Environment and Planning D: Society and Space*, 33(2), 247–264.

Sterne, J. (2003) *The Audible Past: Cultural Origins of Sound Reproduction*. Durham, NC: Duke University Press.

Stiles, K. and Selz P. (2012) *Theories and Documents of Contemporary Art: A Sourcebook of Artists' Writings* (2nd edition, revised and expanded by Kristine Stiles). Berkeley: University of California Press, pp. 470–471.

Thompson, E.A. (2002) *The Soundscape of Modernity: Architectural Acoustics and the Culture of Listening in America, 1900–1933*. Cambridge: MIT Press.

Tsing, A.L. (2015) *The Mushroom at the End of the World: On the Possibility of Life in Capitalist Ruins*. Princeton: Princeton University Press.

Vannini, P. (2015) Non-Representational Research Methodologies: An Introduction. In P. Vannini (ed.), *Non-Representational Methodologies: Re-envisioning Research*. Abingdon: Routledge, pp. 1–18.

Watson, C. (2007) *Weather Report* [CD]. Touch.

Whatmore, S. (2006) Materialist Returns: Practising Cultural Geography in and for a More-Than-Human World. *Cultural Geographies*, 13(4), 600–609.

Yusoff, K. (2014) Geologic Subjects: Nonhuman Origins, Geomorphic Aesthetics and the Art of Becoming Inhuman. *Cultural Geographies*, 22(3), 1–25.

Chapter 10

And all at once the clouds descend, shed tears that never seem to end

Looking from the early modern age at water in the Anthropocene

Simon Meisch

Introduction

In dealing with creative approaches to understanding human–water relationships, this chapter proposes two perspectives on creativity: writing and reading. It will analyse how authors artfully create images of water and readers creatively both co-produce this textual world. Two questions are at the centre of this analysis: How do early modern German poets aesthetically construct water in the context of the climatic changes during the Little Ice Age (1400–1850)? Why might it still be relevant for the inhabitants of the Anthropocene to engage with these texts?

As a strategy to endure the 'global crisis of the seventeenth century' of which the Little Ice Age was part and parcel, Parker (2013, 112) explains how several German 'Lutheran pastors composed hymns that reproached God for 'holding back the sunshine and sending heavy rain'. I will look at three of these pastors (Paul Gerhardt, Simon Dach and Johann Rist) and study four poems that deal with water-related extreme weather events (heavy precipitation, floods, drought and frost). My analysis will situate the aesthetic construction of water within the context of a particular kind of crisis management at a particular time and in a particular place (a religious crisis in German Lutheran territories after 1600) and according to a particular literary genre (Lutheran songs of piety, *Frömmigkeitslied*). This approach thus contextualies water knowledge in times of climate change. It follows Sheila Jasanoff, who discussed how scientific facts concerning the natural world 'detached knowledge from meaning' (Jasanoff 2010, 233). In her view,

> [r]epresentations of the natural world attain stability and persuasive power
> [...] not through forcible detachment from context, but through constant, mutually sustaining interactions between our senses of the *is* and the *ought*: of how things are and how they should be.
>
> (Jasanoff 2010, 236)

One might admit that this is an interesting humanistic endeavour in its own right – and nevertheless ask what these early modern poems really have to tell us about the 'unfolding water drama in the Anthropocene' caused by global climate change and population growth (Rockström et al. 2014).

Against this background, I will first study how four early modern German poems deal with climate-induced, water-related extreme weather and explain these events within a context of meaning. After introducing the authors and poems, I will briefly discuss in what sense these poets speak about climate change and how they represent extreme weather events. As literature is about crisis representation as well as crisis management, I will study how these poets gave meaning to the extreme weather events and what it means that they do so within the literary genre of Lutheran songs of piety. Second, I will ask why it might still be interesting today to read these poems. In this regard, the meaning of reading becomes relevant because it is an act that involves readers, challenges their imagination and can thus affect their perceptions of reality. Finally, this study will be situated within the wider debate on the role of the humanities within integrated, solution-oriented and policy-relevant research called for to deal with the challenges of the Anthropocene.

Writing during the Little Ice Age

The authors and the Lutheran songs of piety

Paul Gerhardt, Simon Dach and Johann Rist were Lutherans and had an academic background in Protestant theology. Paul Gerhardt (1607–1676) worked as a tutor, preacher and pastor in the Berlin region. Today, he is considered one of the most famous German poets of the seventeenth century and the most influential German hymn writer besides Martin Luther (Kemper 1987, 266–272). Johann Rist (1607–1667) made a triple career as pastor near Hamburg, a poet of religious and non-religious poems and a physician and naturalist. Just like Dach, he actively and prominently participated in poetological debates that would eventually establish German as a language of literature. During his lifetime, he was famous for his non-religious literature, whereas today he is considered to be one the most important Protestant hymn writers (Kemper 2006, 172–184). Simon Dach (1605–1659) made a career as professor of poetry and rector of the University of Königberg. Until the mid-twentieth century, Dach's work was held in high esteem, though today he is not widely read (Kemper 2006, 80–89).

In their hymns, the three poets deal with the whole range of hardships of the great crises in the seventeenth century (war, famine, pestilence, etc.). This chapter focuses solely on water-related extreme weather events. In particular, it will look at four hymns that speak about heavy precipitation, wetness and droughts due to frost or long-overdue rain (see Table 10.1).

The songs can be placed within the literary genre of 'Lutheran songs of piety' (*Lutherisches Frömmigkeitslied*). It emerged as part of a movement of

Table 10.1 Early modern songs analysed in this chapter[1]

Author	Song	Sung to the tune of
P. Gerhardt	*Buß- und Betgesang bei unzeitiger Nässe und betrübtem Gewitter* (Song of penitence and prayer in case of unseasonable wet and distressed thunderstorm)	*Wenn wir in höchsten Nöten sein* (When in the Hour of Utmost Need) by Paul Eber (1511–1569)
J. Rist	*Das Sechste Buhßlied. Sehr nützlich zu singen, Wen etwan grosse langwirige Hitze, Und gahr dürre Zeit einfellt* (The sixth song of penitence. Very useful to sing in times of great long heat and drought)	*Kommt her zu mir, spricht Gottes Sohn* (Come to Me, Says God's Son) by Georg Grünewald (c.1490–1530)
J. Rist	*Frommer Haußvätter und Haußmütter andäch-tiges Bittlied zu Gott, Wen es ohne unterlaß regnet und sich die Wasser hefftig ergiessen* (For religious housefathers and house-mothers, a pious song of petition to God when it is raining without cease and the waters pour heavily)	*Wo Gott der Herr nicht bey uns helt* (Where God the Lord Stands with Us Not) by Justus Jonas (1493–1555)
S. Dach	*Buß- und Beth-Lied. Simon Dachen von der kalter Winters-Zeit, Anno 1643* (Song of penitence and prayer. Simon Dach on the cold winter-time in the year 1643)	*Es ist gewißlich an der Zeit* (The Day Is Surely Drawing Near) by Bartholomäus Ringwaldt (1530–1599)

religious discontent within in the Protestant church after 1600. To many of its members, Luther's reformation had come to a standstill. They criticised that the Protestant church was occupied with institutional affairs and issues of dogma but disregarded the renewal of religious life. Thus, outside the church orthodoxy but still within the church, a new type of devotional literature emerged and as part of this, the songs of piety. Originally, these songs were produced for domestic use and family prayers; collections of these songs were hugely popular. Only later, they were incorporated into official hymnbooks (Kemper 1987, 227–265).

After 1600, the number of these Protestant hymns grew considerably (Kemper 1987; Veit 2005). Apart from denominational questions, they increasingly addressed everyday life issues such as famine, war, pestilence – or the weather. Among this rapidly increasing corpus of texts, the number of 'weather songs' grew exceptionally fast (Veit 2005). Some songs were set to music, e.g. the ones by Rist or Gerhardt. However, most of them could also be sung to known tunes of older songs, which provided another (emotional) layer of meaning. For instance, the abovementioned songs of Gerhardt and

Rist were sung to tunes of famous hymns of consolation, while Dach chose the tune of a hymn about the imminent Last Day (cf. Table 10.1).

Early modern climate change: the Little Ice Age

By claiming that Gerhardt, Dach and Rist address climate change in the context of the Little Ice Age (cf. e.g. Lehmann 2005; Parker 2013; Veit 2005), we make presumptions that have to be discussed first (cf. Bühler 2016, 85). Obviously, the poets did not relate their work to the concept of the Little Ice Age, which only emerged in the twentieth century. This concept refers to a period of deteriorating climate in the Northern hemisphere between 1400 and 1850. Although the mean temperature was cooler than before and after, there was no uniform spatial and temporal effect: Some regions suffered more than others and no year resembled the following one (Glaser 2008; Parker 2013). In addition, this was also a period of religious tensions, wars and refugees, hunger and starvation, diseases, energy crises and the like, which is why Parker (2013) speaks of 'the global crisis in the seventeenth century'. Even if we acknowledge that the poets did not discuss the concept of the Little Ice Age, we still have good reasons to believe that their works dealt with its climatic consequences. First, the biographical features of the three poets make it very likely that they were personally affected by these changes. During their lifetime, they witnessed the transition between two main phases of the Little Ice Age in Central Europe. Second, the poems explicitly refer to well-documented extreme weather events that historical climatology attributes to the Little Ice Age (Glaser 2008, 93), such as Dach's song 'On the Cold Wintertime in the Year 1643'.

What applies to the concept of the Little Ice Age holds true for the concept of climate change. Again, the standard definition of climate – the average weather in a specific area over a period of 30 years – was not familiar to seventeenth-century contemporaries. When this chapter speaks about climate change, it follows Hulme who explains that

> [climate] may also be apprehended more intuitively, as a tacit idea held in the human mind or in social memory of what the weather of a place 'should be' at a certain time of year. But however defined, formally or tacitly, it is our *sense* of climate that establishes certain expectations about the atmosphere's performance. The idea of climate cultivates the possibility of a stable psychological life and of meaningful human action in the world. Put simply, climate allows humans to live culturally with their weather.
>
> (Hulme 2015, 3, italics in original)

Accordingly, the poets were aware that the weather was not as it should be. We find this most distinctly articulated by Dach who mourns that heaven did not stand as it used to be but was in disharmony with land and sea. Yet, how do we know that the extreme weathers in the poems were caused by climate change? We do not. Hulme (2014) labels this question as the 'extreme weather blame question' and elaborates its many theoretical presuppositions and methodological challenges. For the context of this analysis, it is only relevant that the climatic changes of the Little Ice Age made extreme weather events more likely and that the poets were aware that their climate (in the sense of Hulme) had changed.

Water-related extreme weather events and agrarian societies

In the event of instable climate, early modern agrarian societies quickly stumbled into a crisis (cf. Glaser 2008, 195–196). The selected poems deal with water-related extreme weathers either by too much (unseasonable wet, ceaseless rain) or too little water (drought, frost). People suffered from its effects directly (e.g. through flooding) or indirectly (e.g. through crop failures). The poems report about the worries and challenges of agrarian societies. Under extreme weather conditions, peasants cannot till the land and crop failures are looming. Wild and farm animals neither find food nor get fed and suffer under heat or cold. Due to frost, fish and birds cannot be caught and sold; due to heat, they perish. Frost causes energy crises because people have to heat more and find dry wood. Heat increases the danger of (forest) fire, heavy rain that of flooding. In his sixth song of penitence, Rist empathically and expressively describes the severe effects of a drought on humans and animals: Peasants are grieving a lot and vignerons are crying more and more.[2] Due to the heat, wildlife is screaming and staring miserably and the fish are blowing up; they all feel their death coming.[3] On top of this, people suffered under corrupt and failing state authorities and wars. These crises particularly affected the poorest. In this respect, Gerhardt is the most articulate author, as he explicitly mentions the failure of social organisation to provide social justice: State authorities are only concerned with their hatred and envy and constantly waging war.[4] Poor people get no peace and are prematurely chased into the grave.[5] The poems also deal with the psychological effects of these climatic changes. They speak of darkness, uncertainty, anxiety, destitution and agony. In Gerhardt's poem, all elements reach out their hands and fear (*angst*) comes from the depth of the sea and the height of the sky.[6]

Water, then, is the resource agrarian societies desperately need in fair amounts and in due time for a steady pace of daily life. At the same time, water turns into an existential threat if it comes abundantly or not at all. This is the moment when the 'extreme weather blame question' arises and people start asking for an explanation and a way out of their misery.

The penal-theological function of water

We can read the poems as historical sources that tell us how early modern people suffered from climate change. However, that would not be enough:

literature does never simply represent the crisis of an extra-textual reality. It creates perceptions of this critical reality in the first place and can be seen as a reaction to and coping with this crisis (Braungart 1999, 443; Bühler 2016, 85). By verbalising extreme weather, the poems make them communicatively available. With this, they can be seen as collective attempts to cope with anxiety, in particular as they were sung in groups (Gamper 2013; Veit 2005) – and in line with magical thinking, to persuade, appease or even 'force' god to restore a harmonic God–human relationship (Böhme 2000).

The three Lutheran poets perceived a climate crisis and by giving it a poetical language explained it. What we read is their interpretation – according to the genre rules of Lutheran songs of piety. The poems' aesthetical construction of the climate crisis is the historical configuration of an older and more general line of understanding natural disasters, the so-called *penal theology*. According to this, extreme weather events are the consequences of human sins and the legitimate punishment by a deity. These events are symptoms of a disrupted God–human communication and as such, a 'semantic event' that needs symbolical heeling by e.g. atonement and repentance (Böhme 2014).

The particular penal theological interpretations within the selected poems use water as an organising idea. First, water is a medium of communication between God and humans. Dach and Rist describe the damages caused by the extreme weathers (frost, drought) with images such as 'heaven as iron', 'earth as brass' or 'powder and dust for rain'. This metaphorical language refers to the Old Testament where God promises his blessing in case people conform to his laws, and threatens with his curses in case they do not (Leviticus 26; Deuteronomy 28). In the former case, rain is promised for human flourishing and in the latter water shortages to 'punish them seven times more for their sins' (Leviticus 26:18). By referring to this imagery, Dach and Rist make water a communicative element in the God–human relationship.

Second, God is seen as the master of the waters – as for instance Rist's 'pious song of petition to God' explicitly states: 'you also set water its aim'.[7] With this, two aspects become apparent. First, water-related extreme weathers demonstrate God's omnipotence. Human sins triggered the punishment but God commands nature and the elements. Second, as God is still active in the creation (*creatio continua*), people can read his will in the Book of Nature and, by implication, in water flows, too (Kemper 1987, 34–65).

Third, water expresses an emotional bond between God and his people. We see this most clearly in Gerhardt's poem. There, the ceaseless rain represents God's countless tears because he is both furious at and sad over the human sins: 'And all at once the clouds descend, Shed tears that never seem to end' (Gerhardt 1867, 295).[8] Meanwhile, the dark and hostile world is also an image of the rotten inner state of (Protestant) Christianity. Therefore: 'sons of man' are called to weep too in their attempt to repent: 'Ah, child of man! go weep alone, Thy many grievous sins bemoan,/Henceforward from thy crimes refrain,/Repent, and be thou clean again' (Gerhardt 1867, 295).[9] In a similar way, Dach uses the bitter frost as an image how the mutual affections of God and his people have been frozen.

Finally, in the poems, water symbolises a crisis of faith. In line with other Lutheran writers of their time, the three poets were afraid that God might exaggerate his punishment and drive people away from faith. The constant divine judgment might undermine the Lutheran dogma that Christ once for all suffered the death on the cross and thereby reconciled God and humans (Kemper 1987, 235). Rist mourns and complains that God lets 'decent people' suffer because their 'dear food' rots in the wetness and that it is now high time to show that people can also expect help from him.[10]

In the poems, we thus witness a dialectical relationship to water, which 'takes on meanings by virtue of its social circumstances, while people's interactions with meaningful water also co-constitute human identities and imaginaries (Linton and Budds 2014, 174).

Reading in the Anthropocene

So, what can the inhabitants of the Anthropocene take away by reading these early modern poems? In a way, this question is badly put because, first, we cannot speak of *the* humans in the Anthropocene[11] and, second, our understanding of reading forbids homogenising assumptions.[12] Reading is a creative act. Each reader plays (more or less consciously) an active part in aesthetically co-producing the textual world (Berendes 2005; Weimar 1999). In order to understand a text, we need to bring in real-world knowledge and different concepts of action and behaviour. Apart from that, by reading we adapt our expectations to the text and modify our own understandings of appropriate action and behaviour. Of course, reading can fail when we lack certain knowledge about the real world, codes of conduct or literary genres or we cannot adapt to the text for whatever reasons. With this, I want to make two points. First, we have good reasons to believe that the textual world will be different for each reader respectively. How one reacts to a text – such as a seventeenth-century hymn – depends on individual and cultural factors or the interaction of both. Second and more importantly, as we are involved in co-producing the textual world, reading does have real-world implications (Berendes 2005). Hofer (2012, 246) explains how conscious reading slows down and intensifies our perceptive abilities, allows us to update and compare different perspectives and worldviews and promotes our self-awareness as readers and our role as attentive observers of the world. Thus, we cannot answer the question of what a particular person takes away when reading the early modern poems. However, in view of discussions about water in the Anthropocene, we can say that we get a space of reflection on important and relevant issues. In the following, five points of view will be outlined.

First, readers experience water as an aesthetic creation. In the poems, it does not only come as representation of material water but also as an emotionally charged bond between God and humans. The creative design of both followed the rules of a particular literary genre. Thus, readers experience representations of water and water-related extreme weather as deeply embedded in *one* specific literary-historical context of meaning. Second and related to this, the successful or failed co-production of water through the readers confronts them with the question of how their *own* waters are (aesthetically) created. After all, water comes in many shapes and homogenising water runs against our intuitions. Third, these poems allow reflection on religious lines of argumentation on water, extreme weather and climate change. Mainstream engagement with these issues takes place within secular techno-scientific framings of today (Benessia et al. 2012). However, we might ask to what extent these mainstream water sciences (implicitly or explicitly) assume the place of religion (Böhme 2014; Linton and Budds 2014). Even in a less radical view, we might ask which religious undertones survived in and how they shape present perspectives of water. Yet, it would be premature to assume *one* secular world because religion plays a (probably increasing) role in all parts of the world. We cannot ignore this when dealing with representations of nature such as water (c.f. Donner 2011; Hochschild 2016; Peppard 2014).

To some and not the least prominent scholars in the field of water research, water turbulences are rising and a water drama is unfolding in the Anthropocene. They argue that humans have altered 'rainfall stability, due to both land-use changes and climate change' (Rockström et al. 2014, 1249). Therefore, in order to secure future water supply and thus human prosperity, they call for a 'shift in thinking about water' and urgent actions (Rockström et al. 2014, 1259). Their suggestions take the form of techno-science and managerial politics of the Earth system. Against this background, the poems offer at least two additional insights. Fourth, apparently just like us today, people of the seventeenth century believed that they caused climate change by social misconduct. While today we tend to naturalise the conflict and look to the natural sciences for techno-scientific solutions, early modern contemporaries regarded it as a social issue. With this, I do not deny anthropocentric causes of climate change. Yet, engaging with this poetry provokes thought about what it means that we *cause* global environmental changes (Hailwood 2016) and if extreme weathers are external to our social organisation (Meisch 2018). Finally, reading these poems invites reflection on doomsday narratives such as the water drama in the Anthropocene. This creates a sense of immediate urgency and might call for extra-ordinary political measures. However, what does this imply for democratic decision-making (Benessia et al. 2012; Rudiak-Gould 2013)? In spite of (physical, social and normative) complexities, we need to conceive the future as open and subject to change by human action (Lövbrand et al. 2015; Swyngedouw 2010).

Meeting the humanities

The approach of this chapter touches on the role of humanities within the sustainability sciences and their potential contribution to solution-oriented, policy-relevant research. Water science is a good example. Humanities are called to leave their comfort zones and participate in this kind of research. With good reasons, humanities scholars remain suspicious (Hulme 2011; Rockoff and Meisch 2015). Present debate on sustainable water takes place within the modernist frame of science (cf. e.g. Rockström et al. 2014) that relates to both the practice of science itself as well as its alignment with particular notions of social progress. First of all, it concerns the scientific quest for abstraction, i.e. for forms of knowledge that can be expressed preferably in mathematical terms and travel through time, space and scale but that become detached from its contexts of meaning. Meanwhile, this abstract knowledge claims to be the exclusive or at least the most relevant form of knowledge about our natural world (Jasanoff 2010). The hydrological cycle is an apt example. Having emerged within a particular historical techno-scientific context, since then it became so predominant that it is now regarded as the 'natural' way of looking at water – despite its obvious normative, cultural, political, etc. omissions (Linton and Budds 2014). The hydrological cycle points to another implication of modernist science, one that preached control over nature and promised universal welfare by providing secure knowledge for techno-scientific and marketable innovations (Benessia et al. 2012). A good case in point is the so-called water, energy and food security nexus claiming the integration of these three sectors and techno-scientific progress under the pretext of human security, especially for the poorest of the poor (Leese and Meisch 2015).

Due to the prevailing narrative of modernist science, scholars in the humanities remain sceptical about calls for an integrative, solution-oriented and policy-relevant science. They challenge the role of science in solving social issues, the focus on problem-solving itself, post-politicisation or the concomitant reduction of sustainable development to neoliberal practices (Benessia et al. 2012; Hulme 2011; Lövbrand et al. 2015; Swyngedouw 2010). Humanities scholars hesitate to contribute to solutions within integrated sustainability science based on 'a homogenizing and reductive simplification of the normative complexity of our environmental situation' (Hailwood 2016, 60).

Against this background, we should not feel bad that the approach in this chapter does not produce abstract and applicable knowledge to solve present water problems. Instead, it argues that reading – for instance, seventeenth-century poems, has other merits (Hofer 2012, 246–247; Weimar 1999). First, it opens reflexive spaces that increase contingency awareness. As literature is an ambiguous interplay of different speech forms, it invites us to observe their interaction, to deal with complex, provisional and uncertain knowledge and to recognise and deliberate our own way of coping with ambiguity, complexity and uncertainty. With this, we have the chance to build an inner attitude to encounter the different social natures of water. For instance, the poems allow observation of how water is aesthetically constructed and embedded in different meaning contexts. Second, it promotes creativity understood as the disposition to recognise and compare alternatives. In a way, by recognising the historical and cultural situatedness of these Lutheran songs of piety, we today make an experience of otherness. Meanwhile, engaging with these poems challenges us to reflect our approaches and concepts on water and water-related extreme weather events. The texts can also question our notions of doomsday narratives and stimulate intellectual engagement with how it would be to live in a world that will always remain beyond our control. Finally, reading imparts our capacity for empathy. For one thing, we get involved with the textual world and are touched by the fortunes and misfortunes of the characters we encounter. On a more general level, reading is always 'talking to oneself in a foreign name' (Weimar 1999). We engage cognitively and emotionally with a world set up by another person. In the case of the seventeenth-century poems, it is to a certain degree an alien and distant world with which we can still emphasise. We can make an experience that might change our perception of how other cultures approach water. In the end, we are invited to contemplate ethically on how water is and should be. When dealing with representations of our natural world, the humanities provide their support by keeping literary knowledge comprehensible and readable. Doing so, they pass on the wealth of human imagination and orienting knowledge and make it accessible to all of us.

Conclusion

This chapter asked how three seventeenth-century Lutheran poets aesthetically created water and meanings surrounding them, and why reading their poems is still relevant today. By doing so, we first learnt how water was co-produced by different forms of knowledge: the personal experience of climate-induced extreme weather events, general experiences of agrarian societies and biblical language within the genre of Lutheran songs of piety. They resulted from a religious crisis in particular German territories after 1600. Their specific image of water emerged between time-bound developments in the context of the great crisis of the seventeenth century (Little Ice Age, wars, religious conflicts, etc.) and images and issues that are timeless, e.g. the weather gods and the 'extreme weather blame question'. This chapter dealt with the creativity of authors but also with the creative act of reading. Readers co-create spaces of (self-)reflection on different representations of their natural world, i.e. in our case, water and water-related extreme weather events. Finally, this chapter situated its reflections on the aesthetical creation of water by authors and readers within the wider debate on the role of humanities within transformative, solution-oriented and policy-relevant (water) science. By doing so, it expressed a sceptical view on the integration of humanities into this scientific enterprise.

Water knowledge is inevitably diverse, socially embedded and historically contingent (Linton 2014; Linton and Budds 2014). After all, there might not be the *one* relevant knowledge for policy solutions but many hybrid ones (Benessia et al. 2012). Literature (as much as other forms of art) provides neither a simple nor the only authoritative representation of nature. However, it can generate and represent specific literary knowledge of human–water relationships. These relationships are embedded in wider moral contexts. Literature thus offers a space for reflection on how we as humans want to live together on planet Earth and deal with a complex world that confronts us with harmful events. Engaging with literature therefore allows us to liberate ourselves from 'modern, standardizing narratives of techno-scientific power and control' (Benessia et al. 2012, 76). We can learn to creatively deal with a complex world we might never fully understand (Taleb 2012) and build 'relative resilience, defined as the capacity to embrace change and complexity and creatively adapt to them, as they unfold' (Benessia et al. 2012, 87).

Notes

1   Gerhardt's poem is printed in Fischer (1906, 435–436); Dach's in Fischer 1906, 84–85; and Rist's 'Frommer Haußvätter und Haußmütter andächtiges Bittlied zu Gott' in Fischer 1905, 276. Rist's 'Das Sechste Bußlied' is taken from Rist 2013, 105–109. If not stated otherwise the translations are made by me.
2   'All Akkersleüte trauren sehr / Die Wintzer heülen immer mehr und mehr' (Rist 2013, 107).
3   'Es schreien auch die wilden Thier / In dieser Dürre für und für / Sie nahen sich dem Sterben/ Sie stehn und gaffen jämmerlich / Die Fisch im Wasser blähen sich / Sie fühlen Ihr Verderben' (Rist 2013, 107).
4   'Man zanckt noch immer fort und fort / Es bleibet Krieg an allem Ort / In allen Winkeln Haß und Neid / In allen Ständen Streitigkeit' (Fischer 1906, 436).
5   'Man plagt und jagt die armen Leut / Eh' als es Zeit, zur Gruben zu / Und gönnet ihnen keine Ruh' (Fischer 1906, 436).
6   'Drum strecken auch all Element / Hier wider uns aus ihre Händ' / Angst kommt uns aus der Tief und See / Angst kommt uns aus der Luft und Höh' (Fischer 1906, 436).
7   'Der du dem Wasser auch sein Ziel / Gesetzet' (Fischer 1905, 276).
8   'Die Wolcken giessen allzumal / Die Trähnen ohne Maaß und zahl' (Fischer 1906, 436).
9   'Ach! wein' auch du, o Menschenkind / Und traure über deine Sünd; Halt doch von deinen Lastern ein / Und mache dich durch Busse rein' (Fischer 1906, 436).
10 'Des Himmels stäte Feuchtigkeit / Läst unsre Saat verderben / Es mus in dieser Ernde Zeit / Die liebe Frucht ersterben / So suchet Gott die Menschen heim,/ Die fleißig sind, aus Koht und Leim/ Die Nahrung zu erwerben. [...] Steh auff, O GOtt, und wende dich / Zu hören unser Flehen / Hilff deinen Kindern gnädiglich / Laß einmahl stille stehen / Den Regen, der ohn' Unterlaß / Verschwemmet das Getreid' und Graß / Daß wir dein' Hülffe sehen' (Fischer 1905, 276).
11 Actually, it is the homogenising feature of the Anthropocene imaginary that by simply referring to humanity ignores social diversity and difference. Lövbrand

et al. (2015, 213–214) call this the 'post-social ontology of the Anthropocene' (cf. also Hailwood 2016; Meisch 2016).

12 Besides, it would be an empirical question beyond the scope of this chapter what people in the twenty-first century (which people?) really take along by reading our early modern poems.

References

Benessia, A., Funtowicz, S., Bradshaw, G., Ferri, F., Ráez-Luna, E.F. and Medina, C.P. (2012) Hydridizing Sustainability: Towards a New Praxis for the Present Human Predicament. *Sustainability Science*, 7, 75–89.

Berendes, J. (2005) Literatur und Moral, Literaturwissenschaft und Ethik. In Maring, M. (ed.), *Ethisch-Philosophisches Grundlagenstudium 2. Ein Projektbuch*, Münster: Lit, pp. 69–83.

Böhme, H. (2000) Anthropologie der Vier Elemente. In Kunst- und Ausstellungshalle der Bundesrepublik (Hg.), *Wasser*. Köln: Druck- und Verlagshaus Wienand, pp. 17–38.

Böhme, H. (2014) Postkatastrophische Bewältigungsformen von Flutkatastrophen seit der Antike. *Zeitschrift für Literaturwissenschaft und Linguistik*, 44(173), 75–93.

Braungart, G. (1999) Poetische Selbstbehauptung. Zur ästhetischen Krisenbewältigung in der deutschen Lyrik des 17. Jahrhunderts. In Jakubowski-Tiessen, M. (ed.), *Krisen des 17. Jahrhunderts. Interdisziplinäre Perspektiven*, Göttingen: Vandenhoeck & Ruprecht, pp. 43–57.

Bühler, B. (2016) *Ecocriticism. Grundlagen – Theorien – Interpretationen.* Stuttgart: Metzler.

Donner, S. (2011) Making the Climate a Part of the Human World. *Bulletin of the American Meteorological Society*, October, 1297–1302.

Donner, S. (2007) Domain of the Gods: An Editorial Essay. *Climatic Change*, 85, 231–236.

Fischer, A. (1904–1916) *Das evangelische Kirchenlied des 17. Jahrhunderts.* Vollendet und herausgegeben von W. Tümpel. Vol. 2 (1905), Vol. 3 (1906). Gütersloh: Bertelsmann.

Gamper, M. (2013) Der Mensch und sein Wetter: Meteo-Anthropologie nach 1750. *Zeitschrift für Literatur*, 1, 79–97.

Gerhardt, Paul (1867) *Paul Gerhardt's Spiritual Songs.* Translated by John Kelly. London: Alexander Strahan.

Glaser, R. (2008) *Klimageschichte Mitteleuropas. 1200 Jahre Wetter, Klima, Katastrophen.* Darmstadt: WBG.

Hailwood, S. (2016) Anthropocene: Delusion, Celebration and Concern. In Pattberg, P. and Zelli, F. (eds), *Environmental Politics and Governance in the Anthropocene: Institutions and Legitimacy in a Complex World*, London: Routledge, pp. 47–61.

Hochschild, A. (2016) *Strangers in Their Own Land: Anger and Mourning on the American Right.* New York: New Press.

Hofer, S. (2012) Literaturwissenschaft und nachhaltige Entwicklung. In Studierendeninitiative Greening the University, Tübingen (ed.), *Wissenschaft für nachhaltige Entwicklung! Multiperspektivische Beiträge zu einer verantwortungsbewussten Wissenschaft*. Marburg: Metropolis, pp. 225–251.

Hulme, M. (2011) Meet the Humanities. *Nature Climate Change*, 1, 177–179.

Hulme, M. (2014) Attributing Weather Extremes to 'Climate Change': A Review. *Progress in Physical Geography*, 38(4), 499–511.

Hulme, M. (2015) Climate and Its Changes: A Cultural Appraisal. *Geography and Environment*, 2, 1–11.

Jasanoff, S. (2010) A New Climate for Society. *Theory, Culture & Society*, 27(2–3), 233–253.

Kemper, H.-G. (1987) *Deutsche Lyrik der frühen Neuzeit. Vol. 2 Konfessionalismus.* Tübingen: Niemeyer.

Kemper, H.-G. (2006) *Deutsche Lyrik der frühen Neuzeit. Vol. 4/II Barock-Humanismus: Liebeslyrik.* Tübingen: Niemeyer.

Leese, M. and Meisch, S. (2015) Securitising Sustainability? Questioning the 'Water, Energy and Food-Security Nexus'. *Water Alternatives*, 8(1), 695–709.

Lehmann, H. (2005) Die Wolken gießen allzumal/die Tränen ohne Maß und Zahl. Paul Gerhardts Lied zur 'Kleinen Eiszeit'. In Behringer, W., Lehmann, H. and Pfister, C. (eds), *Kulturelle Konsequenzen der 'Kleinen Eiszeit'.* Göttingen: Vandenhoeck and Ruprecht, pp. 215–221.

Linton, J. (2014) Modern Water and Its Discontents: A History of Hydrosocial Renewal. *WIREs Water*, 1, 111–120.

Linton, J. and Budds, J. (2014) The Hydrosocial Cycle: Defining and Mobilizing a Relational-Dialectical Approach to Water. *Geoforum*, 57, 170–180.

Lövbrand, E. et al. (2015) Who Speaks for the Future of Earth? How Critical Social Science Can Extend the Conversation on the Anthropocene. *Global Environmental Change*, 32, 211–218.

Meisch, S. (2016) Fair Distribution in the Anthropocene: Towards a Normative Conception of Sustainable Development. In Pattberg, P. and Zelli, F. (eds), *Environmental Politics and Governance in the Anthropocene: Institutions and Legitimacy in a Complex World.* London: Routledge, pp. 62–78.

Meisch, S. (2018) Dem Extremen Sinn und Sprache geben: Extremwetter in den Geistlichen Liedern von Dach, Rist und Gerhardt. In Meisch, S. and Hofer, S. (eds), *Extremwetter. Konstellationen des Klimawandels in der Literatur der frühen Neuzeit.* Baden-Baden: Nomos, pp. 97–121.

Parker, G. (2013) *Global Crisis: War, Climate Change and Catastrophe in the Seventeenth Century.* New Haven, London: Yale University Press.

Peppard, C. (2014) *Just Water: Theology, Ethics and the Global Water Crisis.* Maryknoll, NY: Orbis Books.

Rist, J. (2013) *Neue Himmlische Lieder (1651).* Kritisch herausgegeben und kommentiert von J.A. Steiger. Berlin: Akademie Verlag.

Rockoff, M. and Meisch, S. (2015) Climate Change in Early Modern Literature: Which Place for Humanities in the Sustainability Sciences? In Meisch, S. et al. (eds), *Ethics of Science in the Research for Sustainable Development.* Baden-Baden: Nomos, pp. 269–298.

Rockström, J., Falkenmark, M., Allan, T., Folke, C., Gordon, L., Jägerskog, A., Kummu, M., Lannerstad, M., Meybeck, M., Molden, D., Postel, S., Savenije, H. H.G., Svedin, U., Turton, A. and Varis, O. (2014) The Unfolding Water Drama in the Anthropocene: Towards a Resilience-Based Perspective on Water for Global Sustainability. *Ecohydrology*, 7, 1249–1261.

Rudiak-Gould, P. (2013) We Have Seen It with Our Own Eyes: Why We Disagree about Climate Change Visibility. *Weather, Climate, and Society*, 5, 120–132.

Swyngedouw, E. (2010) Apocalypse Forever? Post-Political Populism and the Spectre of Climate Change. *Theory, Culture & Society*, 27(2–3), 213–232.

Taleb, N. (2012) *Antifragile: Things That Gain From Disorder*. New York: Random House.

Veit, P. (2005) Gerechter Gott, Wo Will Es Hin/Mit Diesen Kalten Zeiten? Witterung, Not und Frömmigkeit im Evangelischen Kirchenlied. In Behringer, W., Lehmann, H. and Pfister, C. (eds), *Kulturelle Konsequenzen der 'Kleinen Eiszeit'*. Göttingen: Vandenhoeck & Ruprecht, pp. 283–310.

Weimar, K. (1999) Lesen: zu sich selbst sprechen in fremdem Namen. In Bosse, H. and Renner, U. (ed.), *Literaturwissenschaft. Einführung in ein Sprachspiel*. Freiburg: Rombach, pp. 49–62.

Part IV

When water disrupts

Water as agent and co-constitutor
of place and culture

Chapter 11

'Water mafia' politics and unruly informality in Delhi's unauthorised colonies

Matt Birkinshaw

Introduction

In this chapter I illustrate the varied politics arising from different socio-technical mediations between humans and water, specifically the informal economy of water supply in tubewells and tankers, in a large unplanned area of Delhi. As a dense and sprawling collection of unauthorised neighbourhoods on the edge of Delhi, this area is hard for state agencies to control and regulate, and this, in turn, gives rise to further arrangements outside of formal regulation, driven by advantage or necessity. An account of water provision in this environment provides sharper insight to what is often a generalised understanding of water access, infrastructure or informality.[1]

In Sangam Vihar, a large lower-income area at the southern edge of Delhi, unpredictable water supply from tanker trucks and tubewells is part of daily life. Groundwater levels are exploited to the point of exhaustion and local leaders are said to control water supply for political, and financial, gain. The new *Aam Aadmi* (common man) Party (AAP) government plans to reform informal water supply and its political consequences in the area by extending piped supply. However, the resilience of informal water is a continuing challenge. After my main fieldwork period, an AAP politician was jailed and the local party split over accusations that both new and old area representatives were working with the 'water mafia'.

Rather than treating Sangam Vihar as representative of informal settlements in general, as some previous researchers have done, I highlight the specificities of place and history, and the relationship these bear to recently introduced attempts to regulate informal water in the area. This adds more empirical detail to previous policy-focused research, which has presented an incomplete picture of water systems in the area and insufficiently outlined their relationship to party politics (Das Gupta & Puri 2005; Kacker & Joshi 2012; Sheik, Banda, Jha, & Mandelkern 2015).

In India, 'mafias' operate in many spheres; land, real estate, sand, fish, onions and so on. The term implies an assertive-aggressive dominance over the market, with a strategic approach to regulation ('informality') facilitated through the involvement of politicians and state employees. The 'water mafia' politics of the chapter title also refers to the way allegations of corruption and complicity with the water mafia are made between local rivals for political advantage, 'playing politics' in Indian English. While researchers working in other cities have expressed scepticism over the veracity of talk of a water mafia (Björkman 2015; Graham, Desai, & McFarlane 2013), in my field site there clearly are identifiable individuals and practices. At the same time, as other researchers have pointed out for corruption-talk more broadly, the popular imagination of these opaque and fluid informal political economies also suggests something about people's understanding of the local state (Doshi & Ranganathan 2017; Gupta 2012; Ranganathan 2014). However, the informal political economy of water, as a combination of natural and social elements, extends beyond the state and social relations, and is difficult for any actor to sustainably capture and control. It is, in a word, *unruly*.

Unruliness, is not confined to unauthorised colonies or off-grid water arrangements, the concept of informality itself is difficult to subsume under a single rubric. The politics of informal groundwater supply illustrate this by offering an alternative picture to the predominant tendency in work on urban informality to focus on land and labour (e.g. McFarlane & Waibel 2012; Roy & AlSayyad 2004; Schindler 2016). The need to consider urban water infrastructures as social hydrology embedded in a wider ecological context (Malghan, Kemp-Benedict, Goswami, Muddu, & Mehta 2013) forces us to include the materialities of these informal economies, again in a way rarely present for work on other sectors (see Hull 2012). Groundwater, often accessed through decentralised, off-grid infrastructures, presents a different set of governance dynamics to prominent discussions of the politics of piped and informal supply in other Indian cities, which usually focus on the public sector (Anand 2011; Björkman 2015; Coelho 2006; but see Ranganathan 2014). Groundwater is now India's *main* water source across all uses (Cullet 2014). However, despite the rapid depletion of north Indian groundwater (Kulkarni, Shah, & Vijay Shankar 2014), this 'underground political ecology' of water, both sub-soil and illicit, has been under-researched to date (Bebbington 2012; Birkenholtz 2009; Maria 2008; Rohilla 2012).

The fieldwork that I draw on was conducted as PhD research over 21 months between 2013 and 2015, with a follow-up visit in 2017. I lived in my research sites for a majority of the time, 12 months in an urban village and 9 months in an unauthorised colony, both areas with a limited and unreliable connection to the water network. I spent considerable time with participants in their workplaces and everyday surroundings, and met with interviewees on multiple occasions as far as was possible. In Sangam Vihar and Deoli, I spoke to over 100 residents across eight blocks, six municipal wards and both legislative assembly areas, mostly in unauthorised colonies but also in urban villages and clusters of informal huts. I also spoke to local politicians and party workers across parties, water suppliers (tankers, bottles and borewell

networks), real estate dealers, health workers and non-governmental organisation (NGO) workers as well as a large number of people more active at the
city level, including Delhi Water/Jal Board (DJB) staff, politicians, academics,
NGO workers, activists and other researchers. Communication was often in
Hindi, but I have translated to English here, retaining the original in italics if
necessary. All names of people in body text, except the Chief Minister, have
been changed.

Groundwater dependence and urban informality

Delhi is a city of around 17 million people with highly uneven water supply
(Shaban & Sharma 2007). Most residents stay in various types of 'informal'
neighbourhoods outside the Delhi masterplan (Bhan 2009). Government
agencies are unwilling to provide services to these unauthorised areas and
the infrastructure that exists is often self-provided or constructed through
politicians' discretionary funds. Around half the population regularly uses
non-government water (Zérah 2000) and for one in four people the only
government water supply is through tanker trucks (Comptroller and Auditor
General 2009). A major source of additional water is mechanically pumped
groundwater, particularly in peripheral newer areas with less government
water. Delhi sits at the apex of north India's dramatically falling groundwater
and, as in many of India's fastest growing cities, the city's water tables have
sunk drastically in recent years (Kulkarni & Shah 2015).

Sangam Vihar is an agglomeration of 38 unauthorised colonies at Delhi's
south-eastern border. A sprawling settlement of dense low-rise concrete
buildings, the area is said to house around a million people and is divided
into two constituencies represented by elected members of the Delhi state
Legislative Assembly (known as MLAs). In 2016, the area was allowed piped
water supply, however, during most of my fieldwork, Sangam Vihar relied on
water tankers (trucks carrying between 4,000 and 30,000 litres) and tubewells
(a pipe and submersible pump inserted into the underground aquifer, also
known as borewells), operated by public and private suppliers. Politicians,
journalists and local people talk of 'water mafia' involvement in relation to
both tankers and tubewells.

'Off-grid' and informal water supply in Sangam Vihar varies due to several
factors, demonstrating the challenges of theorising informality or infrastructure in the abstract and the necessity of 'disaggregation' in discussions of an
assumed aggregate 'water access' (Ahlers, Cleaver, Rusca, & Schwartz 2014).
The different spatial, temporal and material qualities of 'off-grid' decentralised
informal water, public and private, supplied from tankers (mobile, transitory,
visible, peripatetic) and tubewells (localised, fixed, submerged, constant) lend
themselves to political and social (class, caste, gender) influence in different
ways, and give rise to more specific, varied and politicised forms of governance than formal piped supply. I suggest that these variations in 'informal'

water demand a more nuanced approach than a formal/informal binary (Ahlers et al. 2014; Truelove 2016; also see Allen, Hofmann, Mukherjee, & Walnycki 2016; Tutu & Stoler 2016).

Most households in Sangam Vihar use multiple water sources; irregular supply from tubewells (DJB and private), topped up with tanker water and 20-litre cans for drinking. For a household of five, water bills of over 3,000 rupees (£30) a month is quite likely. The monthly household income for a poor (below poverty line) household that make up one in six of Delhi's population was 3,060 rupees in 2005 (Government of NCT of Delhi 2009, 239).

Many households purchase 20 litre plastic 'cans' of water to drink if they can afford it – even poor migrant labourers, or those living in tarpaulin shacks. Suppliers and DJB staff told me that these cans are borewell water filtered and bottled locally. Depending on their size, households would purchase between one and three locally produced cans a day. Daily wage labourers, the elderly and other people with less money, cannot afford these cans and drink water from tubewells or tankers after boiling it.

The frequency and reliability of public tankers from the Delhi Jal (water) Board (DJB) varies. In some areas DJB tankers come weekly or daily, in others they never do. As residents must be present when the tanker arrives to fill their buckets and tanks, they have to wait for the tanker to arrive and women and children miss school and work. Even tanker owners and drivers admit that arguments and fights break out in the rush to fill buckets. Asha, a local resident, and BJP supporter, described her frustration with the water situation to me.

> The DJB tankers do not always come. They take the water and sell it to someone else they have a 'setting' with or someone they know. Like you remember the tanker I ordered from the [DJB staff member at the local office]? That never came.
> (Interview with local resident [female, c.30], BJP supporter, 17 June 2015)

Politicians and party workers from the BJP and AAP, as well as NGO workers and residents, told me that DJB drivers might sell whole tanker loads of water on the black market. When I was shown Delhi's GPS tanker tracking system there were some readings from tankers way off their scheduled route, which was hard for the operative showing me the system to explain. Several people told me that government tanker drivers also sometimes deliver to the correct location but sell the water, which should be free, to residents by the bucket for 10 or 20 rupees. The amount taken per tanker (roughly 800 rupees) then approximates the black market rate. Presumably the funds from these various black market sales accrue to the actors in control of the tankers; public tankers can be either 'emergency' one-off deliveries that fall under the MLAs management, or scheduled regular deliveries under the DJB. In other cases, it seems tanker drivers sell water on their own account.

Tubewells are the other main water supply method in Sangam Vihar. They provide groundwater by sinking a pipe containing a submersible pump into the underground aquifer. Water is pumped to street-level pipes leading to people's houses. In 2016, there were 165 government tubewells in Sangam Vihar. Using the population data and tubewell yields available this averages one tubewell for every 6,000 residents (Bhardwaj 2015a; Centre for Science and Environment 2010, 4–5). If all wells were running 24 hours a day this would give only 36 litres per person per day. This is below the WHO minimum recommendation of 50 litres for household needs and and poses a risk to residents' health (Gleick 1996; Howard & Bartram 2003).

In addition to government tankers and tubewells, privately owned tankers and tubewells also supply water. There are estimated to be 200 private tubewells in the Sangam Vihar area. They are more expensive than government tubewells, but do not supply water any more regularly; again from once in 10–12 days to once in 6 weeks. Many residents I spoke to were connected to both government and private networks. Prices for private networks varied from 500–800 rupees a month.

Tubewell water is very hard, with a high level of dissolved solids and other impurities. Residents understand that this can make people sick and causes kidney problems (Bellizzi et al. 1999). People with money wouldn't use tubewell water for drinking unless they had no choice; however poor people would not be able to afford other options. The water is dirty at different and unpredictable times and will change colour and 'go bad'[2] if left standing for more than a day (I was shown water that had been left standing with a suspension of brown flakes floating in it). This is obviously a problem if water is only available once every two to four weeks.

The users and operators I spoke with described government tubewells networks ranging in size from 20 houses to '3–500 minimum'. In some blocks government tubewell networks supplied water to each house for an hour a day. However, most networks are oversubscribed, and many people said that the government tubewells gave them water once or twice a month. In areas towards the edge of the colony, supply was less frequent and people might get water once in five to six weeks or two months. For DJB tubewells, households pay monthly fees, officially 50 rupees, but up to 200 may be charged, by the *paani kholnewala* ('water opener') responsible for opening the valves for different lanes and houses. This person was also referred to as the *paani malik* ('water owner') and often appeared to be connected to the area *pradhan*, an informal local leader.

When the area was first settled, residents could find water at 30 feet with handpumps, however now borings at least 650 feet deep are required. Several interviewees mentioned the declining groundwater level. Water extracted from nearer the bottom of the aquifer is increasingly hard, salty and gritty.

This causes the submersible pumps used in tubewells to wear out very fast. One female *kholnewala* I met had quit the job as declining water levels were causing too many people to complain. In other areas the idea of complaining to the (male) *kholnewala* was quite novel, and other *kholnewalas* said that people did not complain (see Kacker & Joshi 2012). Indeed, residents often seemed wary, perhaps even afraid, of some tubewell managers.

Kholnewalas described themselves to me as employees of the DJB junior engineer and said that the DJB paid for the electricity but did not collect revenues. One *kholnewala* told me that: 'in other areas government men would do this work but they cannot in Sangam Vihar' (cf. Ghertner 2017). The role appears to be a pragmatic compromise at the edge of the formal utility. This pragmatism on the part of the DJB may be related to the 'capture' (*kabza*) of government tubewells by powerful people described by water suppliers, residents and party workers from both parties. Several residents, BJP and AAP supporters, said this was done by the dominant castes in the villages who had owned the land are now well represented in local politics and real estate (Kacker & Joshi 2012, 31). 'Villagers capture the tubewells and sell it to people from [other states]. The migrants cannot say anything. The tubewell people are charging 500–600 [rupees] a month' (Interview with tanker owner [male, *c*.45], BJP associate, 16 August 2015).

Residents, researchers and water dealers stated that tubewell capture was conducted with the complicity of the DJB, as well as politicians, local magistrates and police. This illustrates the popular, and academic, understanding of Delhi's 'mafias', in tankers, tubewells or land, as relying on connections spanning across, and blurring the boundaries between, public and private actors and agencies (cf. Ranganathan 2014). Individuals, tankers and water may move positions creatively between legal and illegal, public and private, philanthropy and profiteering. Privately owned water tankers illustrate this as they may deliver water for both DJB beneficiaries and private buyers, both of whom may be paying customers.

Private tankers are more reliable than government tankers, but more expensive. A whole tanker of 4,000 litres costs from 800–2,000 rupees, depending on the supplier, client and time of year. Households will club together to order and share the water. Private tankers fill from both private and public tubewells within Sangam Vihar, as well as outside. Private tankers filling from government tubewells have been reported to other researchers also (Sheik et al. 2015, 5). While most research, in Delhi and other Indian cities, describe local tanker monopolies (e.g. Borthakur 2015; Kjellén & McGranahan 2006; Ranganathan 2014), in Sangam Vihar, both residents and water suppliers told me that there is healthy competition between private tankers and they do not have set territories. Private tankers operations in the area are usually small and appear often to be run as a side business.

Asha, introduced above, told me that in her area the borewell has been captured by the *pradhan* and is used to fill his tanker, which supplies to the

DJB. This means he is using groundwater for official (nominally free) deliveries and selling the treated DJB water to higher-paying customers.

> There is not [enough] water in the [tubewell] pipes but the private tankers are selling it. They take it and sell it anywhere. The people then have to buy private tankers or cans or boil water. You remember Shiv at Masjid Mohalla? There is water in his borewell but he does not give it to the people in the lane at Masjid Mohalla: it is for his people only [. . .] So there is no water in the pipes for the people paying 100 rupees. At this time there is no water, it is difficult.
> (Personal communication, resident [female, *c.*30], BJP supporter, 17 June 2015)

In summary, Sangam Vihar's water supply modes are more varied and complex than a planned part of the city. There are also important differences between 'informal' water supply modes, which differ across locations (e.g. number of users) and within them (e.g. proximity to supply). Both tankers and tubewells are disruptive in their limited provision of infrequent and unpredictable amounts and qualities of water. In addition to uncertainty and stress, the use of private tankers, tubewells and bottles/cans also impose heavy financial, time and health costs on low-income households. Sangam Vihar's hydraulic segregation from the rest of the city rests on a decision of the city water board and city government. With the amounts and qualities of water from tankers and tubewells stretched to the limit by rising populations and falling water tables, the issue is an emotive topic for political campaigning.

Water and politics, disruptive and productive

Water is a big theme in Delhi politics. Inadequate water supply enables a wide range of established and aspiring political entrepreneurs to represent an area and lobby for, or make, improvements. Intercession with higher authorities on behalf of their constituents appears to be a basic activity in the roles of neighbourhood social workers, NGO staff, party workers, local leaders (*pradhans*) and elected politicians (counsellors, MLAs and MPs) alike. The visibility of the day-to-day interactions and labour of politicians and party workers ('political work') is an important element in Indian democratic politics (Anand 2011; M. Banerjee 2010; Björkman 2014, 2015).

In the 2013 Delhi state elections, the new party, AAP, campaigning on anti-corruption and access to urban services, won enough seats to form a minority government. A large part of the AAP's campaign focussed on removing the 'vote bank' politics that left informal neighbourhoods reliant on political patronage to access government services. After 49 days, citing obstruction from Municipal and Union governments, both dominated by the Bharatiya Janata Party (BJP), the AAP government resigned. In the 2014 Delhi elections, the AAP returned to win a landslide victory, taking 67 of the Delhi parliament's 70 seats. The BJP won three seats, and Indian National Congress, the previous ruling party, none. The current city Chief Minister, AAP's Arvind Kejriwal, first gained his reputation as a 'disruptor' by campaigning against water privatisation in 2003 (S. Banerjee 2014). South Delhi's member of the national parliament (MP), a BJP representative, is said to 'have made his money in water', if not his reputation. He is from the dominant caste of one of the local villages, mentioned earlier for their involvement in water, real estate and politics, and has several relatives in south Delhi politics, three at MLA level.

In Sangam Vihar, some people described politicians from all parties as 'being all the same', promising (improved) water in exchange for votes, but 'doing nothing' or being 'in it for the money'. Others attributed these motives to opposition party representatives, with varying degrees of balance. Cases have been registered against both BJP and AAP politicians in disputes connected to the 'water mafia' (ANI News 2016; NNC News 2014).

A second narrative from residents was that corrupt officials prevented improvements to water supply. Asha described the limited water as 'political games'; 'all the politicians are talking about water but they do not give details. They just say 'give us your vote and we'll give you water'. Asha also told me that the DJB official responsible had refused to provide water to the area without a side-payment. A *pradhan* in another neighbourhood gave a similar account:

> The previous MLA didn't get the authorisation to extend DJB water [. . .] because the DJB, the MLA and the [sub-divisional officer] were taking money. Just like the DJB might ask for money for a new connection this was for a whole new area. Tanker businesses are there too [. . .] The MLA would say 'if you want work done you must give me money'.
> (Personal communication, *pradhan, kholnewala* [male, *c*.55], 3 July 2015, F91)

Both water tankers and borewells are said to be closely related to the economy of party politics. I was told by a BJP MLA candidate that control over 'emergency' tankers is understood as one of the perks of MLA office, and may be used as a source of income or means of patronage. Installation of tubewells is arranged by a request to the MLA and patronage is also said to occur in the location and speed of approval for borewell requests. More directly, party supporters simply capture and control tubewells. One of the young volunteers at an AAP MLA office told me that:

> Before us the MLA was BJP [name removed]. He placed his people on the water sources. They consume water and sell water. They were making money from water sources [. . .] They would give or not give – [because] this one is BJP person, this one is not. They would have people on the pipelines, they are strong people in fighting mode so even if one

person gets good water and other gets none they will not say [. . .] It was like this in the whole MLA area.

(Interview, AAP party worker [male, *c*.33], 18 May 2015, F81)

The income from public and private tubewells can be quite substantial and is passed from the local workers to higher up the party to fund election expenses, after a deduction of around 10 per cent as a 'salary' for the local workers – an amount close to the wages of other *kholnewale* I spoke to. Tej, an AAP volunteer, put it bluntly: 'The BJP capture the bore wells and sell the water, it is a mafia'. Other AAP party workers also described this situation. Residents of Sangam Vihar have previously told other researchers that borewell capture is linked to the MLA's office under earlier governments (Bhardwaj 2015b; Sheik et al. 2015). Senior BJP workers too said that tubewells and tankers allow patronage and rent-seeking.

The AAP election campaigns used both water and corruption in public services as key issues. AAP posters stated: 'No water, so no vote (a warning to power)'. This reversed the logic of informal areas voting for politicians on the promise of services, and encouraged people to refuse to vote again for representatives who had not delivered. Tankers and tubewells are over-stretched in Sangam Vihar, and the rapidly declining groundwater levels in the south of the city can only add to this dynamic. Consequently, AAP's political rhetoric challenging the 'water mafia' appealed to voters, and the many residents unregistered to vote, frustrated with the deterioration of urban services (Bhardwaj 2015a).

Once elected, AAP was quick to take visible initiatives against 'water mafia'. Some 11 days into the first AAP government in Delhi, a DJB 'taskforce' and 'a large police force' began 'a massive combing operation' in Sangam Vihar and Deoli. They were briefed to 'takeover illegal borewells' being operated for 'commercial gain' (Indian Express 2014). While the intention was to disrupt black market water sales linked to their political opponents, the 'renationalisation' drive, however, only happened where the party had the local strength to do it. Some 35 borewells were passed over to AAP supporters during the '49 days' of the first AAP government. However, this did bring some tubewell network prices down and this, combined with better control of private and public tankers led to some nostalgia for the first AAP government among people I spoke to.

A young AAP volunteer, Ravi, showed me the tubewell in his lane. It had four pipes, each supplying a lane of houses, and a fifth pipe leading to the BJP supporter's large house. He explained about the *kholnewala* system:

The BJP guy was earlier doing the *kholnewala*. These *kholnewala*'s are just gangsters. The people with power will take the job. But we took over in the 49 days. Now [the BJP guy] is doing it again, but in three [or] four days I will be doing this job.

(Personal communication, AAP volunteer [male, *c*.21], 18 May 2015)

I was not convinced that Ravi was likely to dislodge the BJP *kholnewala* as easily as he said, especially after he told me that he had been attacked by BJP workers while campaigning. Echoing Ravi's disclosure that the tubewell had been 'recaptured', a tanker owner and local resident also commented on the resilience of these local power dynamics: 'There was a big fuss about AAP capturing tubewells in the 49 days. They brought DJB vehicles and the police so people were afraid but slowly-slowly it went back to the same people' (Personal communication, tanker owner and NGO worker [male, 45], BJP associate, 16 August 2015).

As well as tubewells, the AAP government was working to address the 'tanker mafia' too. Staff at the MLA office were tasked with keeping a register of complaints over non-arriving tankers and pursuing these with the DJB. The party workers were able to fine tankers that had not delivered; however, the drivers often claimed to have 'broken down'. The AAP also made data from the GPS trackers in the newer tankers publicly available on the DJB website. If the system is working this shows when a tanker is off-route. Unfortunately, no GPS-equipped tankers work in Sangam Vihar, and my conversations and observations suggest that reform of tubewell and tankers was challenging.

> A large group of women is requesting assistance as they are not getting water. They say that the water, tubewell and tankers, is captured by powerful people who live near a Beauty Parlour – Ritesh knows who they mean. He says all he can do is send an emergency tanker in 3–4 days. When I ask the women how they have been getting water they say 'fighting – with the people that are capturing it'. They also say that they are having to go to other places to bring water back in buckets.
>
> (Observations at MLA office, 18 June 2015)

Some people in the area have said that the AAP people running the tubewells were discriminating against them. Similarly, residents (with BJP connections) have stated that 'government tankers are "controlled" by local AAP volunteers who take money and mostly provide water to their loyalists' (Kapil 2016). BJP supporters, including the MP, have also led protests against an AAP 'water mafia' (The Hindu 2015). A volunteer working closely with AAP in Sangam Vihar said that while collecting fees from tubewells by party workers continued under the AAP government it was 'much less' than before. This accusation of continuing extraction of rents through water has also been made by another party worker and led to a split within Sangam Vihar AAP (Singh 2016). The tensions between idealism and pragmatism was palpable in my conversations with party volunteers and leaders during this time (also see Sharma 2015). The decentralised infrastructures used in Sangam Vihar seem to facilitate more localised governance constellations than fixed-network infrastructure. This appears to illustrate the blurred and porous boundaries of the Indian state that other researchers have observed in rural service provision and urban land use (Benjamin 2004; Gupta 1995). A significant difference in this case, making water a slippery political ally, is that the resource in question is seasonally variable, randomly distributed (a property of the quartzite aquifer) and rapidly being depleted.

In this regard, the most substantial reform from the AAP government is to connect Sangam Vihar to the piped network. This was described by AAP party workers as a 'permanent solution' to the problems of water and corruption in Sangam Vihar. At the time of research this work was underway in several blocks. The government has stated an intention to use the area as a case study for connecting unserved areas (Vatsa 2016). The AAP also passed the groundbreaking *Jal Adhikar* Connection Act 2016 (Right to Water Connection Act), which allows all residents in the capital a legal metred connection regardless of tenure status (Bhan 2016). Visiting in 2017, residents in some areas are now connected to network water that has been released into the main pipes. Lane committees collected contributions and had pipes laid themselves, the MLA is said not to have provided any assistance. Other areas, at higher elevation or those unable to collect money for the work are still without piped water. Piped treated water through individual house connections is without doubt a significant improvement for residents. As to the 'water mafia', from my conversations with water dealers, residents and party workers, it seems likely that informal water suppliers in the area will shift into other businesses or locations.

Conclusions

Groundwater provides an input that partially mitigates dissatisfaction over inadequate government supply, while the cost of providing potable water is passed on to consumers forced to rely on expensive private tankers, tubewell networks and 'local' bottled water (Das Gupta & Puri 2005). Diverse methods of water supply in Sangam Vihar, and the different topographies, temporalities and visibilities of tankers, tubewells and bottled water, lead to far more fractured governance patterns than the piped network (cf. Mol & Law 1994). For example, tankers are mobile and hard to track. Tubewells are locally situated and susceptible to capture by proximate power-holders. Both are convenient sources of economic and political capital. Tankers and bottled water suppliers appear to function more as informal enterprises without strong spatial exclusivity. These decentralised infrastructures appears more porous to local power than networked infrastructure. The water supply methods studied in this research are socially embedded in ways that differ to those shown in research on other cities (cf. Björkman 2015; Borthakur 2015; Kjellén & McGranahan 2006; Ranganathan 2014). The complexity of water governance arrangements in this case suggests that we need to move beyond the binaries of formal/ informal water and governance.

My research suggests that the minimal 'off-grid' government water service through tubewells and tankers allows a greater personal role for local representatives and agency workers, discretionary elements of which are often compounded by rent-seeking and political calculation. Any resulting modifications, and the process itself, can generate partial benefits and introduce deviations into the officially sanctioned patterns of water and influence. This gives rise to the phenomenon of 'water mafias' and in turn an accusatory 'water mafia politics'. I suggest that tensions over water led to the election of the AAP based on their campaign against corruption in urban service delivery. Thus the material qualities of water (aquifer exhaustion) disrupted the water and vote-bank politics hegemonic in the settlement. Against this background, the AAP's programme to extend piped supply to the settlement is intended to disrupt prevailing patterns of water governance. However, this programme, and the local party itself, has been disrupted in turn by 'water mafia politics'. Within the grammar of local politics, inadequate water access is a signifier of political corruption and neglect. Thus although the AAP was able to capitalise on dissatisfaction with water supply in the area, they have only partially been able to provide network supply and the murky qualities of informal water have been easily mobilised against them once in power. The informal political economy of water, as a combination of natural and social elements, extends beyond the state and social relations, and is an unruly informality difficult for any actor to sustainably capture and control. This complexity of physical infrastructure and governance arrangements for water supply networks, and their proximity to local social power holders, is underestimated by reform initiatives, whether public or private sector.

Notes

1 I would like to thank all friends and acquaintances in Sangam Vihar, Deoli, and beyond, who shared their time and understandings with me. I am grateful to Layli Uddin, Katyayni Seth, Tom Cowan, Thomas Crowley, Kate Birkinshaw and an anonymous reviewer for feedback on the draft. Any errors remain my own.
2 Hindi: 'kharab ho jaega' – this phrase was often used.

Bibliography

Ahlers, R., Cleaver, F., Rusca, M., & Schwartz, K. (2014) Informal Space in the Urban Waterscape: Disaggregation and Co-Production of Water Services. *Water Alternatives*, 7(1), 1–14.

Allen, A., Hofmann, P., Mukherjee, J., & Walnycki, A. (2016) Water Trajectories Through Non-Networked Infrastructure: Insights from Peri-Urban Dar es Salaam, Cochabamba and Kolkata. *Urban Research & Practice*, 10(1), 22–42.

Anand, N. (2011) Pressure: The PoliTechnics of Water Supply in Mumbai. *Cultural Anthropology*, 26(4), 542–564.

ANI News. (2016) *FIR Against AAP MLA: Dinesh Mohaniya Says It's an Attempt to Defame Him*. Retrieved 28 February 2017, from www.youtube.com/watch?v=aas7MxFQoMc.

Banerjee, M. (2010) Leadership and Political Work. In P. Price & A.E. Ruud (eds), *Power and Influence in India: Bosses, Lords and Captains*. New Delhi: Routledge India, pp. 20–43.

Banerjee, S. (2014) *The Disrupter: Arvind Kejriwal and the Audacious Rise of the Aam Aadmi*. New Delhi: Rupa Publications.

Bebbington, A. (2012) Underground Political Ecologies: The Second Annual Lecture of the Cultural and Political Ecology Specialty Group of the Association of American Geographers. *Geoforum*, 43(6), 1152–1162.

Bellizzi, V., De Nicola, L., Minutolo, R., et al. (1999) Effects of Water Hardness on Urinary Risk Factors for Kidney Stones in Patients with Idiopathic Nephrolithiasis. *Nephron*, 81(Suppl. 1), 66–70.

Benjamin, S. (2004) Urban Land Transformation for Pro-Poor Economies. *Geoforum*, 35(2), 177–187. https://doi.org/10.1016/j.geoforum.2003.08.004

Bhan, G. (2009) 'This Is No Longer the City I Once Knew': Evictions, the Urban Poor and the Right to the City in Millennial Delhi. *Environment and Urbanization*, 21(1), 127–142.

Bhan, G. (2016, September 2) A City Is for All Its Citizens. *The Hindu*.

Bhardwaj, A. (2015a, January 24) All Parties Love This Illegal Colony: Its People Don't Count, Their Votes Do. *Indian Express*. New Delhi.

Bhardwaj, A. (2015b, January 28) Where Power Flows Through Water Pipelines. *Indian Express*. New Delhi.

Birkenholtz, T. (2009) Groundwater Governmentality: Hegemony and Technologies of Resistance in Rajasthan's (India) Groundwater Governance. *Geographical Journal*, 175(3), 208–220.

Björkman, L. (2014) 'You Can't Buy a Vote': Meanings of Money in a Mumbai Election. *American Ethnologist*, 41(4), 617–634.

Björkman, L. (2015) *Pipe Politics, Contested Waters: Embedded Infrastructures of Millenial Mumbai*. New Delhi: Orient Blackswan.

Borthakur, N. (2015) *Urban Water Access: Formal and Informal Markets: A Case Study of Bengaluru, Karnatka, India* (Working Paper No. 1). Hyderabad: School of Public Policy and Governance, Tata Institute of Social Science.

Centre for Science and Environment. (2010) *CSE Water Audit*. Delhi: Centre for Science and Environment.

Coelho, K. (2006) Tapping In: Leaky Sovereignties and Engineered (Dis)Order in an Urban Water System. In M. Narula, S. Sengupta, R. Sundaram, A. Sharan, J. Bagchi, & G. Lovink (eds), *Sarai Reader 06: Turbulence*. New Delhi: Centre for the Study of Developing Societies, pp. 497–509.

Comptroller and Auditor General. (2009) *Report of the Comptroller and Auditor General of India on Social Services in Delhi*. New Delhi: Comptroller and Auditor General.

Cullet, P. (2014) Groundwater Law In India: Towards a Framework Ensuring Equitable Access and Aquifer Protection. *Journal of Environmental Law*, 26(1), 55–81. https://doi.org/10.1093/jel/eqt031

Das Gupta, P., & Puri, S. (2005) *Private Provision of Public Services in Unauthorised Colonies: A Case Study of Sangam Vihar* (Working Paper). New Delhi: Centre for Civil Society.

Doshi, S., & Ranganathan, M. (2017) Contesting the Unethical City: Land Dispossession and Corruption Narratives in Urban India. *Annals of the American Association of Geographers*, 107(1), 183–199.

Ghertner, D.A. (2017) When Is the State? Topology, Temporality, and the Navigation of Everyday State Space in Delhi. *Annals of the American Association of Geographers*, 107 (3), 731–750.

Gleick, P.H. (1996) Basic Water Requirements for Human Activities: Meeting Basic Needs. *Water International*, 21(2), 83–92. https://doi.org/10.1080/0250806960 8686494

Government of NCT of Delhi. (2009) *Economic Survey of Delhi 2008–2009* (Economic Survey of Delhi). New Delhi: Planning Department, Government of NCT Delhi.

Graham, S., Desai, R., & McFarlane, C. (2013) Water Wars in Mumbai. *Public Culture*, 25(169), 115–141. https://doi.org/10.1215/08992363-1890486

Gupta, A. (1995) Blurred Boundaries: The Discourse of Corruption, the Culture of Politics, and the Imagined State. *American Ethnologist*, 22(2), 375–402. https://doi. org/10.1525/ae.1995.22.2.02a00090

Gupta, A. (2012) *Red Tape: Bureaucracy, Structural Violence, and Poverty in India*. Durham, NC: Duke University Press.

Howard, G., & Bartram, J. (2003) *Domestic Water Quantity, Service Level and Health* (No. WHO/SDE/WSH/03.02) (p. 33). Geneva: World Health Organisation. Retrieved from www.who.int/water_sanitation_health/publications/wsh0302/en/

Hull, M. (2012) *Government of Paper: The Materiality of Bureaucracy in Urban Pakistan*. Berkeley: University of California Press.

The Hindu. (2015, June 17) Protests Left, Right and Centre. *The Hindu*. New Delhi.

Indian Express. (2014, January 9) Delhi Jal Board Task Force Takes on Water Mafia, Illegal Tubewells. *Indian Express*. Delhi.

Kacker, S.D., & Joshi, A. (2012) Pipe Dreams? the Governance of Urban Water Supply in Informal Settlements, New Delhi. *IDS Bulletin*, 43(2), 27–36.

Kapil, S. (2016, June 12) No Water, Just Pipe Dreams. *Deccan Herald*. New Delhi.

Kjellén, M., & McGranahan, G. (2006) *Informal Water Vendors and the Urban Poor*. London: International Institute for Environment and Development (IIED).

Kulkarni, H., & Shah, M. (2015) Urban Water Systems in India. *Economic And Political Weekly*, 50(30), 57–69.

Kulkarni, H., Shah, M., & Vijay Shankar, P.S. (2015) Shaping the Contours of Groundwater Governance in India. *Journal of Hydrology: Regional Studies*, 4(Part A), 172–192. https://doi.org/10.1016/j.ejrh.2014.11.004

McFarlane, C., & Waibel, M. (2012) *Urban Informalities: Reflections on the Formal and Informal*. Farnham; Burlington, VT: Ashgate.

Malghan, D., Kemp-Benedict, E., Goswami, R., Muddu, S., & Mehta, V.K. (2013) Social Ecology of Domestic Water Use in Bangalore. *Economic and Political Weekly*, 48(15), 40–50.

Maria, A. (2008) Urban Water Crisis in Delhi. *IDDRI, Working Papers No.06/2008*, 23.

Mol, A., & Law, J. (1994) Regions, Networks and Fluids: Anaemia and Social Topology. *Social Studies of Science*, 24(4), 641–671.

NNC News. (2014) *Interview with Ramesh Bidhuri 04 04 14*. New Delhi: NNC. Accessed February 28 2017, from www.youtube.com/watch?v=c9AKDm9tZe4.

Ranganathan, M. (2014) 'Mafias' in the Waterscape: Urban Informality and Everyday Public Authority in Bangalore. *Water Alternatives*, 7(1), 89–105.

Rohilla, S.K. (2012) *Water, City and Urban Planning: Assessing the Role of Groundwater in Urban Development and Planning in Delhi* (Working Paper No. id:5068).

Roy, A., & AlSayyad, N. (2004) *Urban Informality: Transnational Perspectives from the Middle East, Latin America, and South Asia*. Lanham, MD: Lexington Books.

Schindler, S. (2016) Seeing and Governing Street Hawkers Like a Fragmented Metropolitan State. In S. Chakravarty & R. Negi (eds), *Space, Planning and Everyday Contestations in Delhi*. New Delhi: Springer, pp. 21–34.

Shaban, A., & Sharma, R.N. (2007) Water Consumption Patterns in Domestic Households in Major Cities. *Economic and Political Weekly*, 42(23), 2190–2197.

Sharma, S. (2015, January 23) The Bhushans Aren't the Only Ones Upset with Arvind Kejriwal. Accessed 14 February 2017, from https://scroll.in/article/702088/the-bhushans-arent-the-only-ones-upset-with-arvind-kejriwal

Sheik, S., Banda, S., Jha, B., & Mandelkern, B. (2015) *Limbo in Sangam Vihar: Delhi's Largest Agglomeration of Unauthorised Colonies* (Cities of Delhi). New Delhi: Centre for Policy Research.

Singh, A. (2016, April 1) Welcome To Sangam Vihar, Where the Most Common Way to Access Water Is Illegal. Accessed 14 February 2017, from www.youthkiawaaz.com/2016/04/water-crisis-delhi-sangam-vihar/

Truelove, Y. (2016) Incongruent Waterworlds: Situating the Everyday Practices and Power of Water in Delhi. *South Asia Multidisciplinary Academic Journal*, 14, 101–103.

Tutu, R.A., & Stoler, J. (2016) Urban But Off the Grid: The Struggle for Water in Two Urban Slums in Greater Accra, Ghana. *African Geographical Review*, 35(3), 212–226.

Vatsa, A. (2016, February 12) We Are Confident that Delhi Will Have Its Best Summer this Year: Kapil Mishra. *Indian Express*. New Delhi.

Zérah, M.-H. (2000) *Water, Unreliable Supply in Delhi*. New Delhi: Manohar and Centre de Sciences Humaines.

Chapter 12

Encountering water

Sensitivities and practices for moving beyond 'Big Water' interventions[1]

Claire Hoolohan and Alison L. Browne

Introduction

Accounts of the co-evolution of systems of provision with everyday practices are well established, and existing research has illustrated the intersections between ordinary patterns of water use and the infrastructures and institutions that supply water in domestic spaces. Several studies have demonstrated how encounters between people and water within the Big Water system – that which provides access to an expansive and uniform water supply and service in several developed nations – influence and normalise intensive patterns of everyday water use, thus contribute to such challenges as water scarcity, drought and ecosystem degradation (e.g. Sofoulis 2005; Taylor and Trentmann 2011). However, despite this literature offering valuable insights, researchers elsewhere are calling for broader recognition of the multiple relationships between water and society; accentuating how forms of encounter that exist alongside modern infrastructural systems differentially shape everyday geographies (Bear and Bull 2011). Our aim in this chapter is to respond to this call; to explore how encounters with water beyond the 'Big Water' system shape everyday practice, and reflect on what this means for strategic responses to socio-ecological crises. We begin with a brief synopsis of the relationship between everyday water use and Big Water. The analysis that follows considers two sites of alternative encounter – participatory river restoration schemes and camping music festivals – and examines the opportunities that such differential experiences of water afford for the ongoing emergence of everyday practice.

Big Water and the prevailing hydrosocial contract

Big Water is the term used by Sofoulis (2005) to describe the centralised provision of water services in nations such as Australia, the USA and the UK that established a near universal penetration of domestic water supply during the twentieth century. In the UK, a long history of structural engineering has resulted in an expansive water supply and sewerage system that provides for more than 99 per cent of the population (DWI 2016). This expansive physical infrastructure is supported by centralised regulation such that, aside from negligible regional differences in taste and texture, the water flowing through the taps in one home is near indistinguishable from that in any other.

Big Water not only describes the infrastructures of supply, but a certain hydrosocial contract that is expressed in institutional arrangements, cultural norms, regulatory frameworks and the built environment (Farrelly and Brown 2014). A hydrosocial contract describes the implicit understandings of water management arrangements that, though unwritten, legitimise certain modes of water governance (Turton and Meissner 2000). Big Water is characteristic of a governance system that aspires to 'abstract waters of the earth from the chorological and cultural contexts that otherwise give them meaning' (Linton 2010, 104) and produces, purifies and commodifies water so as to render it a cooperative resource (Kaika 2005). This contract is enacted and reinforced through the historical and ongoing development of water supply systems, in which successive governments have employed logics and techniques of hydraulic engineering to deliver a wholesome and reliable supply of water and ensure its effective removal (Bakker 2003). Over time, tacit understandings of the distribution of responsibility for water resources emerge that support such technocratic development, and become self-perpetuating in light of the continued successes of the Big Water system in supplying and removing water from the home (Linton 2010).

As well as managerial arrangements, Big Water has implications for the meanings and practices associated with water use in domestic spaces. As Allon and Sofoulis (2006, 48) describe:

> Big Water's collective conventions of water use and distributions of responsibility for it are 'baked in' to domestic objects, including standard water fittings like taps, drains, sewer systems and automatic washing machines: convenient 'user-friendly' interfaces designed to make it easy to use water.

Repetitive engagement with Big Water during everyday routines thus diffuses into the social imaginary of water services to shape ideas regarding what normal and necessary water use looks like (Sofoulis 2005). Further, as a taken-for-granted component of modern urban planning, Big Water becomes engrained in the design of technologies and homes, and subsequently determines the options and possibilities for how water needs are fulfilled (Kuijer and de Jong 2011; Yates and Evans 2016). In these ways our encounters with Big Water shape domestic demand by influencing social conventions and aiding the steady integration of water in everyday rituals of cleanliness, comfort and convenience (Shove 2003). Thus, Big Water acts as an inconspicuous facilitator; enabling and reinforcing the increasingly intensive patterns of water use that contribute to regional supply-demand deficits (Chappells and Medd 2008), and restricting personal and institutional reflection on how water services might otherwise be attained (Browne et al. 2014; Hoolohan 2016).

Hydrosocial encounters beyond Big Water

Despite their prevalence, Big Water systems are not the only site of encounter between people and water. Throughout their lives, people engage in spaces where water persists in more or less altered forms (e.g. rivers, lakes or canals). There are also occasions in which the absence of Big Water is conspicuous (e.g. in the presence of alternative water supply systems, or when water services are not available as is accustomed (Sofoulis 2014; Woelfle-Erskine 2015); and incidents in which the parameters of Big Water are exceeded (e.g. during drought (Chappells et al. 2011) or flood (Whatmore 2013)). These various opportunities are geographically discontinuous, with uneven patterns of access and affect, and more or less voluntary forms of participation. However each disrupts the continuity of Big Water systems, enabling people to see, sense and interact with water not wholly incorporated in the Big Water system (Dicks 2014).

We suggest that attempts to understand the relationships between systems of provision and everyday patterns of water use have, to date, largely discounted people's encounters with water beyond the Big Water system. However their potential to unsettle hydrosocial contracts and espouse novel modes of interaction between people and water make them worthy of further consideration. In the remainder of this chapter we investigate how such encounters with water affect people's experiential understandings of water services, the social exchanges that circulate around water use and embodied understanding of how water is used. We reflect on the opportunities such encounters pose for shaping the ongoing emergence of patterns of water use.

In order to examine these aims, we draw on findings from two independent studies. The first focuses on *Care for the Kennet*, a strategic programme of participatory river restoration projects designed to counteract the historical abstraction of domestic water from its natural entanglements. These reflections are drawn from research towards a PhD, involving interviews, focus groups and observational data (cf. Hoolohan 2016; Hoolohan and Browne 2016). The second set of reflections considers camping music festivals as an incidence in which access to Big Water is impeded, therefore presenting an opportunity for novel interactions between people and water (see case study 2). The reflections are drawn from 60 qualitative interviews and 265 quantitative surveys across two UK festivals (cf. Hitchings et al. 2018).

Case study 1: Care for the Kennet

In European water management, public participation in conservation, restoration and resource management is reinforced via mechanisms such as the Water Framework Directive (WFD) that calls for the active inclusion of interested parties and the public. Common examples of public participation include environmental quality monitoring, river restoration, wildlife monitoring and domestic water conservation (Paloniemi et al. 2015). The example discussed in this chapter, *Care for the Kennet*, is illustrative of such activities, and demonstrates the role of local interest groups, charities and trusts in coordinating these initiatives.

Care for the Kennet is a community-led initiative that aims to 'reconnect the water in people's homes to the water in the river' (Spokesperson at ARK AGM 2012). A partnership between local ecological action group, Action for the River Kennet (ARK) and the regional water company, Thames Water, *Care for the Kennet* comprised of a programme of activities designed to enable local residents to interact with the river from which their water is supplied. Several of these activities targeted children and their parents, providing opportunities to engage with the river and learn about the ecosystem that it supports. For example, facilitated by ARK, children learned to care for trout eggs, young European eels and mayflies, releasing them into the Kennet after 8–10 weeks. Other activities included river visits and interactive stalls at community events designed to establish a broader understanding of the connections between water and society in the local community.

The literature on environmental education highlights the importance of such personal engagements in local environments to facilitate 'deep learning'; tacit and intuitive understandings of environmental sustainability that support problem solving (Lloyd and Gray 2014). Consistent with the literature in support of public participation in environmental management, evaluation of *Care for the Kennet* described how the initiative facilitates social learning, thereby increasing the efficacy of water conservation efforts (Hoolohan 2016). However, the connections between such learning, hydrosocial contracts and everyday practices are under-researched, and therefore the opportunities they present for sustainability intervention is poorly understood.

Case study 2: camping music festivals

The UK live music industry – including the ubiquitous camping festival – continues to grow rapidly. Made famous by the iconic Glastonbury festival, camping music festivals take place all over the UK from spring to early autumn. Festivalgoers, typically of a younger age group (Mintel 2015), camp for an extended four- or five-day weekend. Although there are differences in the facilities provided at different festivals, water and energy provisioning is usually limited with portable toilets providing basic sanitation, drinking water provided through standpipes or water stations and limited showering facilities. Few other water services are typically available to festivalgoers, though infrastructures for water provisioning are becoming increasingly diversified and commoditised (Hitchings et al. 2018).

In recent years the rise of 'glamping' areas often affords access to water-flushing toilets, warm showers and electricity that allows festivalgoers to use hairdryers and straighteners. Further companies are increasingly buying spaces within the festival site to offer additional services, such as a more sanitary toilet experience, a hot shower or a jacuzzi or sauna with friends. These services are typically accessible to a minority of festivalgoers and for a fee, thus the general festival experience is one of infrastructural disruption (water, energy, food), and pleasurable distraction (music, friends, culture, alcohol) (Hitchings et al. 2018).

An increasing body of research is giving greater consideration to festivals as spaces for sustainability education, including to explore how they provoke different engagements with nature (Gibson and Wong 2011; Zifkos 2015). However, recent research shows that festivals are also spaces in which people's normal experiences of cleanliness and hygiene services are disrupted, and where new cultural conventions can be temporarily established (Hitchings et al. 2018). Festivals therefore undo the idea that cleanliness and hygiene cultures are fixed and disconnects the provision of these services from the Big Water system.

Encountering water, developing alternative hydrosocial sensitivities and practices

In both of the case studies, people's interactions with water and water supply systems differ from their ordinary, everyday encounters with Big Water. This difference is a source of potential change; unsettling familiar expectations, conventions and materialities and creating space for the reconfiguration of practices of water use. However, to date there has been limited research to examine the outcomes of encounters beyond the Big Water system. In the following sections we illustrate how, through these encounters, arise alternative experiential understandings of water, novel social exchanges and new memories and therein propose that these encounters possess the potential to influence the ongoing evolution of patterns of ordinary consumption.

Shaping experiential understandings of water services

During *Care for the Kennet* it was observed that encounters between local residents and the rivers from which their water was supplied served to repopulate the public imaginary with other, non-human water users. Alternative understandings of water services emerged as local residents were immersed in activities designed to enhance the ecological functions of the Kennet. These understandings were empathic towards the services that water provides to river species and their ecosystems (Hinchliffe and Whatmore 2006), and recognised the connections between personal water use and the river's continued capacity to provide these services to others. Activities tailored towards children made these connections most explicit, with games and lessons structured around the monitoring of river species thereby drawing attention to their dependency of the river as a shared natural resource. Other activities adapted a similar narrative for different audiences, propagating a broader appreciation for interdependencies between society and water. Thus, by catalysing and legitimising novel forms of interaction with water outside the home, *Care for the Kennet* contributes to tacit understandings of how river life is affected by water supply and use in the local area. These interactions provide a valuable counter-narrative to Big Water's anthropocentric framing of water services, which strips water of its socio-natural context (Kaika 2004).

The camping music festival provides a different encounter beyond Big Water that nevertheless unsettles the assumptions contained within modern water management. Festivalgoers experience a lack of access to technologies of water supply meaning that ordinary water-using practices related to cleanliness, comfort and hygiene are reconfigured or put on hold. In some cases, festivalgoers replaced their morning or evening ritual with other practices – a wet wipe wash, or hair wash under a standpipe in the festival field – but in many cases these routines were suspended altogether. This embodiment of disrupted service provokes the temporary reorganisation of how people manage the body and the self in the absence of water (and energy) infrastructures, thereby untangling water use from Big Water's material systems.

Thus, despite being starkly different, both of these encounters challenge the unwritten assumptions embedded in Big Water's hydrosocial contract, making conspicuous the taken-for-granted articulations of water's function-ality and untangling water services from material systems of supply.

Creating conversations: speaking the unspoken qualities of water

As well as shaping individuals' understandings of water services, novel forms of social exchange arise around these encounters with water beyond the Big Water system. At festivals, the enjoyment of indolence was not an instant embrace but a gradual, emergent legitimisation of 'not doing' rendered acceptable by collective recognition that normality was too difficult to achieve in the constrained infrastructural context (Hitchings et al. 2018). Thus, experimentation in cleanliness and hygiene routines at festivals should be viewed not as individual, but collective and social. This is evident in the 'talk' that emerged at the festival. The conspicuous disruption – to infrastructure, practices and routines – elicits conversation regarding topics of personal cleanliness and hygiene that are typically 'taboo', or at least beyond what is typical of everyday conversation (Browne 2016). The diversity of everyday practice – the varying standards of cleanliness people hold and the multitude of ways people have of performing hygiene and personal care routines – is usually hidden in private spaces of the home. Yet the festival experience enables such differences to be observed and discussed with a greater level of tolerance and co-operation than exists outside the festival, where there are pressures to maintain 'civil' and 'polished' personal hygiene practices (Hitchings et al. 2018). Therein festivals provide spaces in which the limits and preconceived boundaries of acceptable bodily conduct might be scrutinised; allowing new social norms about what is acceptable in personal hygiene to emerge (within the context of the festival).

Social exchanges also play a role in *Care for the Kennet*. The hands-on methods used throughout the initiative create interactions between people and water that exceed those typical in everyday routine, extending the typical geographies of interaction with water and the nature of embodied engagement. Consequently these interactions are remarkable and spark discussion both between participants, and with others. These discussions extend the impact of the initiative beyond the immediate participants, and beyond the life of the project. For example, the invention of games that incorporated the identification and monitoring of river species, establishing the basis for a 'running commentary' (female, age 42) that extend appreciation for river ecologies within family and peer networks. Similarly, river visits and other restoration activities were located close to local amenities (e.g. shops, schools, playgrounds and paths), providing a talking point for the wider community to engage with as they went about everyday activity. Thus the interaction between people and the river create space for broader discussions, such as regarding the boundaries of agency and responsibility for water management. Some of this discussion was critical, highlighting the challenges and contradictions of Big Water that were revealed in the degraded health of local watercourses. In other instances, discussions were solution-oriented; alluding to alternative configurations of water services systems and management practices that might better enable healthy rivers; and domestic practices that might alleviate more acute crises (such as dry rivers and point source pollution incident).

Thus, by creating space for discussions that are atypical of those arise everyday, encounters with water beyond the Big Water system enhance people's appreciation for difference, both of other people's practices, and the needs of other water users. Further, these social exchanges provide a space to critique, negotiate and invent alternative practices of pertaining to water use, water management and bodily conduct.

Creating memories and capabilities

The previous section illustrates how encounters with water beyond Big Water challenge modern hydrosocial contracts. Yet in both cases the continuation of these sensitivities and emergent practices is questionable, as habituated practices of water use are typically reinstated once re-immersed in the Big Water system. Nonetheless, we argue that the residual memories and capabilities derived from these encounters are important, filtering into the ongoing emergence of everyday practice and enhancing people's latent capacity to adapt to future disruption.

In the case of the festival, people described a reversion to pre-existing routines once they returned home (Hitchings et al. 2018). However, whether or not people 'snap back' to their normal ways of doing things when they reconnect with 'real life' is less significant than observing how these alternative encounters establish alternative sensitivities and capabilities that might gradually infuse in the ongoing emergence of domestic routines. Experiencing the festival enables a greater personal appreciation for the diversities of ways in which other people perform their routines of cleanliness and hygiene. The need to navigate alternative material arrangements creates new practical understandings about ways of maintaining personal hygiene without the associated water use. Both of these processes enable reflection on personal and social thresholds to dirt and sweat (e.g. the point at which hair 'needs' washing and bodies begin to smell). These processes contribute to a stock of practice memories (Maller and Strengers 2013), and a widened set of alternative skills and knowledge-abilities of doing things differently, or flexibly. In turn, these memories extend people's capabilities to responding to future infrastructural disruptions, for example as a result of climate change, or other change (Gibson et al. 2015; Head et al. 2016).

Similarly, once away from the river and back within domestic spaces participants in *Care for the Kennet* described a return to ordinary water use, and the critical narratives introduced in encounters with the river were subdued by the continued presence of Big Water. However, the emergent appreciation of the multi-functionality of water resources persists and begins to unsettle practices that were previously performed without critical reflection. The heightened appreciation for the ecosystem services that water provides creates friction for familiar conventions of everyday water use that co-evolve in Big Water systems. Participants reported a heightened appreciation for the variable conditions of the river, rendering periods of low rainfall and incidents such as pollution (e.g. BBC 2013) meaningful and affective. Yet the continuation of water in the household, unresponsive to such incidences, instilled confidence in normal modes of conduct. Thus the sensitivities and practices that emerge within the space of the initiative are not forgotten, but become contested and conflicted. As a result, participants reported contradictory expectations about what was appropriate in terms of water use: to use water to fit with the conventions of everyday water use, but also to find more convivial modes of consumption that resonate better with the appreciation for nonhuman others derived from encounter with the river.

Discussion

As is the case in several Western developed nations, the Big Water system is largely successful in delivering a steady supply of high-quality water to buildings throughout the UK. Thus, it is easy for academic enquiry to fixate on Big Water systems and their effects, not least due to their physical grandeur and ubiquity. However these case studies demonstrate that there are spaces where alternative hydrosocial relationships exist alongside Big Water systems. These two case studies provide illustrative points rather than exhaustive examples, yet combined they demonstrate how the hydrosocial contracts arising beyond Big Water contradict the conventions embedded in pervasive socio-technical systems and therefore possess the potential to unsettle patterns of ordinary water use. The first case study demonstrates how participatory ecological conservation projects, like *Care for the Kennet*, might propagate ecological sensitivities that emphasise the interdependencies between people and water, reframing the expectations of water systems in more-than-human terms. The second example illustrates how temporary suspense of infrastructures and routine, in this instance in a music festival, exposes previously taken-for-granted assumptions about ordinary water use, and provides opportunities for social learning that might support the diffusion of alternative practices of water use.

However, increasing pressures on water supply systems resulting from population growth and climate change challenge the ongoing resilience of large infrastructural systems across the globe. Simultaneously, a growing number of actors from academic, policy and industry spaces are highlighting the importance of consumption, such that global water management challenges are thought not to require a re-emergence of water megaprojects to mitigate risk, but substantive changes to patterns of everyday water use. Thus, the observed significance of these alternative spaces of encounter for experiential understandings, social exchange and practical memory leads us to question what it might mean to take such experiential pluralism seriously, and how we might learn from these examples to inform future sustainability interventions. It is to these questions the final section of this chapter turns.

Demand management has emerged as a core component of water industry activities. Yet despite recent developments (see Russell and Fielding 2010 for a review) there is a substantial and growing critique that highlights an over-reliance of sustainability interventions on simplistic models of consumption (Browne 2015). The perpetuation of techno-economic models of demand, that overemphasise efficiency measures and behaviour change, neglect the relational dimensions of water use, and dismiss the materiality of everyday life. Subsequently, though current permutations of demand management present low-risk solutions for impending environmental crises, the scope of intervention remains limited (Hoolohan 2016). One notable omission, given the scope of this chapter, is any exploration of how a more ecologically beneficial hydrosocial contract might be supported through the substantive reconfiguration of socio-technological systems.

These critiques of water interventions are connected to a wider body of critical research that challenges the current framing of sustainability interventions (e.g. Geels et al. 2015; Strengers 2011). However, to date, there has been little cross-fertilisation of this literature with research on the new ways of managing water in urban spaces or wider perspectives on sustainable urbanisation (e.g. Dicks 2014; Farrelly and Brown 2011). These wider literatures – supported by the examples reflected upon in this chapter – suggest that interventions targeting individuals and households without consideration of the broader hydrosocial contract embedded in urban design will be ineffective in bringing about substantial or long-term changes to domestic demand. Instead what is called for are adaptive frameworks that prioritise 'flexible, inclusive, and collaborative practices, operating within organisational cultures that embrace experimentation' (Farrelly and Brown 2011, 721) in order to develop the water service systems required to address emerging environmental challenges. Thus, rather than technological and behavioural interventions, these findings suggest a need for interventions that might cultivate more convivial ecological sensitivities, capabilities and practices (Hoolohan 2016).

The combination of these discussions repositions demand as emergent from the relationship between the performance of everyday practices that consume water and the enactment of the broader hydrosocial contract. Subsequently, the continuity of Big Water systems becomes framed as part of the problem, perpetuating expectations of water services that are incompatible with the long-term resilience of socio-natural systems, and replicating familiar geographies of responsibility for water management led by water management experts with consumers responsible for moderating personal water use (Allon and Sofoulis 2006). New approaches are being articulated that move away from the responsibilisation of the consumer and their practices as the sole location of change (Evans et al. 2017), and instead position the body, the home and the community as sites for experimentation, where there are opportunities for transformative encounters with water.

The case studies in this chapter begin to illustrate that even for consumers living with Big Water systems, spaces of encounter with water are multiple. We demonstrate how these encounters, not just those with the prevailing Big Water system, have implications for the ongoing evolution of experiences, expectations and capabilities from which emerge practices of everyday water use. Thus, though the examples presented in this chapter may seem marginal compared to the scale of the potential water crises faced both in the UK and worldwide, the findings nonetheless demonstrate the importance of embodied experience and participation in sustainability initiatives. Participation, be it in strategic initiatives, such as *Care for the Kennet*, or occasional spaces in which water infrastructure is reconfigured, such as festivals, provides opportunities to interrogate previous assumptions and taken-for-granted understandings of water, and to develop sensitivities and practices that better reflect the ecologies of water, both human and more-than-human (Hinchliffe and Whatmore 2006). The findings also illustrate the value of conversational dialogue in creating a social imaginary of water that accommodates – and encourages appreciation of – difference (i.e. different needs of people, and also of ecological systems).

The case studies presented in this chapter call for greater critical reflection on how interventions designed to influence everyday water use create, reinforce or marginalise various hydrosocial relations and consequently have implications for the ongoing emergence of water-related practices. In analysing examples of the camping music festival and a participatory ecological restoration project, we demonstrate the potential embodied in encounters with water beyond Big Water to lead to alternatively configured social imaginaries of water and capabilities for water use. Engagement in these spaces is, however, uneven. Uneven spatial distribution of rivers and festivals, combined with unequal geographies of access and participation, mediate the affective capacity of such encounters. Further, strategic efforts to create alternative encounters in order to alter the trajectory of water-using practices are fairly uncommon. *Care for the Kennet* is an unusual example in having made explicit connections between ecological restoration and domestic practice. The growing importance of finding means of steering everyday consumption towards less-intensive patterns of resource use means that understanding how to maximise participation in these spaces, amplify their beneficial effects and further understand their consequences is vital (see also Davies and Doyle 2015).

Conclusion

In this chapter we have shown that interventions into water futures would benefit from being participatory and immersive, engaging as much in the materialities of water supply and demand, as in the social. Consequently, novel physical infrastructures, such as rainwater harvesting schemes (Ward and Butler 2016), community gardens (van Holstein 2015), and greywater recycling (Woelfle-Erskine 2009) become sites of demand management intervention, providing spaces to encounter water that spark new experiential understandings, social exchanges and memories and in doing so the opportunity to enhance the adaptive capacity of domestic practices (cf. Gibson et al. 2015; Maller and Strengers 2013). Our reflections in this chapter also have implications for urban planning more generally, calling for greater recognition of how opportunities for productive encounters with water are mediated in the built environment and everyday geographies of leisure and mobility. This research is therefore also a call to develop connections between research on everyday consumption and that on water sensitive cities (Bell 2013; Ferguson et al. 2013) or urban daylighting (Wild et al. 2011) and similar. Finally, the social science research community needs to articulate more strongly the ways in which we can intervene in the hydrosocial contract in a way that influences both the systems (of provision that usually provide water services) and the practices (of the everyday) that underpin water resource use. In doing so we might be able to imagine water futures where adaptive capacity is not only reflected across infrastructures and natural systems but also into people's everyday embodied skills and capacities.

Note

1 Research in this chapter has been funded by an ESRC Case PhD Studentship Award co-funded by Thames Water through the NWDTC ESRC CASE doctoral scholarship (ES/J500094/1); with the Festivals project funded by the ESRC 'Patterns of Water' project (RES-597-25-003) and the UCL Bridging the Gap Fund in collaboration with Tullia Jack (Lund) and Russell Hitchings (UCL).

References

Allon, F. and Sofoulis, Z., 2006. Everyday Water: Cultures in Transition. *Australian Geographer*, 37 (1), 45–55.

Bakker, K., 2003. *An Uncooperative Commodity: Privatizing Water in England and Wales*. Oxford: Oxford University Press.

BBC, 2013. River Kennet Pollution: Pesticide Blamed for River Fly Death [Online]. *BBC Online*. Available from: www.bbc.co.uk/news/uk-england-wiltshire-23177777 (Accessed 7 July 2014).

Bear, C. and Bull, J., 2011. Guest Editorial. *Environment and Behaviour*, 43, 2261–2266.

Bell, S.J., 2013. Creating Sustainable Urban Water Systems. *Institution of Civil Engineers. Proceedings. Urban Design and Planning*, 166 (DP2).

Browne, A.L., 2015. Insights from the Everyday: Implications of Reframing the Governance of Water Supply and Demand from 'People' to 'Practice'. *Wiley Interdisciplinary Reviews: Water*, 2 (4), 415–424.

Browne, A.L., 2016. Can People Talk Together About Their Practices? Focus Groups, Humour and the Sensitive Dynamics of Cleanliness in Everyday Life. *Area*, 48 (2), 198–205.

Browne, A.L., Medd, W., Pullinger, M. and Anderson, B., 2014. Distributed Demand and the Sociology of Water Efficiency. In K. Adeyeye, ed. *Water Efficiency in Buildings: Theory and Practice*. Chichester, UK: John Wiley & Sons, 74–84.

Chappells, H. and Medd, W., 2008. From Big Solutions to Small Practices. *Social Alternatives*, 27 (3), 44–49.

Chappells, H., Medd, W. and Shove, E., 2011. Disruption and Change: Drought and the Inconspicuous Dynamics of Garden Lives. *Social & Cultural Geography*, 12 (7), 701–715.

Davies, A.R. and Doyle, R., 2015. Transforming Housheold Consumption: From Backcasting to HomeLab Experiments. *Annals of the Association of American Geographers*, 105 (2), 425–436.

Dicks, H., 2014. A Phenomenological Approach to Water in the City: Towards A Policy of Letting Water Appear. *Environment and Planning D: Society and Space*, 32 (3), 417–432.

DWI, 2016. Private Water Supplies in England and Wales [Online]. Available from: www.dwi.gov.uk/private-water-supply/index.htm [Accessed 14 January 2017].

Evans, D., Welch, D. and Swaffield, J., 2017. Constructing and Mobilizing 'the Consumer': Responsibility, Consumption and the Politics of Sustainability. *Environment and Planning A*, 49 (6), 1396–1412.

Farrelly, M.A. and Brown, R.R., 2011. Rethinking Urban Water Management: Experimentation as a Way Forward? *Global Environmental Change*, 21 (2), 721–732.

Farrelly, M.A. and Brown, R.R., 2014. Making the Implicit, Explicit: Time for Renegotiating the Urban Water Supply Hydrosocial Contract? *Urban Water Journal*, 11 (5), 392–404.

Ferguson, B.C., Frantzeskaki, N. and Brown, R.R., 2013. A Strategic Program for Transitioning to A Water Sensitive City. *Landscape and Urban Planning*, 117, 32–45.

Geels, F.W., McMeekin, A., Mylan, J. and Southerton, D., 2015. A Critical Appraisal of Sustainable Consumption and Production Research: The Reformist, Revolutionary and Reconfiguration Positions. *Global Environmental Change*, 34, 1–12.

Gibson, C., Head, L. and Carr, C., 2015. From Incremental Change to Radical Disjuncture: Rethinking Everyday Household Sustainability Practices as Survival Skills. *Annals of the Association of American Geographers*, 105 (2), 416–424.

Gibson, C. and Wong, C., 2011. Greening Rural Festivals: Ecology, Sustainability and Human–Nature Relations. In C. Gibson and J. Connell, eds. *Festival Places: Revitalising Rural Australia*. Bristol: Channel View Publications, 92–105.

Head, L., Gibson, C., Gill, N., Carr, C. and Waitt, G., 2016. A Meta-Ethnography to Synthesise Household Cultural Research for Climate Change Response. *Local Environment*, 21 (12), 1467–1481.

Hinchliffe, S. and Whatmore, S., 2006. Living Cities: Towards a Politics of Conviviality. *Science as Culture*, 15 (2), 123–138.

Hitchings, R., Browne, A.L. and Jack, T., 2018. Should There Be More Showers at the Summer Music Festival? Studying the Contextual Dependence of Resource Consuming Conventions and Lessons for Sustainable Tourism. *Journal of Sustainable Tourism*, 26 (3), 496–514, doi: https://doi.org/10.1080/09669582.2017.1360316.

Hoolohan, C., 2016. *Reframing Water Efficiency: Towards Interventions that Reconfigure the Shared and Collective Aspects of Everyday Water Use*. Thesis submitted to the University of Manchester for the degree of Doctor of Philosophy (PhD) in the Faculty of Science and Engineering.

Hoolohan, C. and Browne, A., 2016. Reframing Water Efficiency: Determining Collective Approaches to Change Water Use in the Home. *British Journal of Environment & Climate Change*, 6 (3), 179–191.

Kaika, M., 2004. Interrogating the Geographies of the Familiar: Domesticating Nature and Constructing the Autonomy of the Modern Home. *International Journal of Urban and Regional Research*, 28 (2), 265–286.

Kaika, M., 2005. *City of Flows: Modernity, Nature, and the City*. New York: Routledge.

Kuijer, L. and de Jong, A., 2011. Practice Theory and Human-Centred Design: A Sustainable Bathing Example. In T. Härkäsalmi, I. Koskinen, R. Mazé, B. Matthews and J.-J. Lee, eds. *Nordes 2011*. Helsinki, Finland: School of Art and Design, Aalto University, 221–227.

Linton, J., 2010. *What Is Water? The History of a Modern Abstraction*. Vancouver: UBC Press.

Lloyd, A. and Gray, T., 2014. Place-Based Outdoor Learning and Environmental Sustainability within Australian Primary School. *Journal of Sustainability Education*, September, 1–15.

Maller, C. and Strengers, Y., 2013. The Global Migration of Everyday Life: Investigating the Practice Memories of Australian Migrants. *Geoforum*, 44, 243–252.

Mintel, 2015. *Music Concerts and Festivals: UK August 2015* London. Mintel Group.

Paloniemi, R., Apostolopoulou, E., Cent, J., Bormpoudakis, D., Scott, A., Grodzińska-Jurczak, M., Tzanopoulos, J., Koivulehto, M., Pietryk-Kaszyńska, A. and Pantis, J. D., 2015. Public Participation and Environmental Justice in Biodiversity Governance in Finland, Greece, Poland and the UK. *Environmental Policy and Governance*, 25 (5), 330–342.

Russell, S. and Fielding, K., 2010. Water Demand Management Research: A Psychological Perspective. *Water Resources Research*, 46 (5), 1–12.

Shove, E., 2003. *Comfort, Cleanliness and Convenience: The Social Organization of Normality*. Oxford: Berg.

Sofoulis, Z., 2005. Big Water, Everyday Water: A Sociotechnical Perspective. *Continuum*, 19 (4), 445–463.

Sofoulis, Z., 2014. The Trouble with Tanks: Unsettling Dominant Australian Urban Water Management Paradigms. *Local Environment*, 20 (5), 529–547.

Strengers, Y., 2011. Beyond Demand Management: Co-Managing Energy and Water Practices with Australian Households. *Policy Studies*, 32 (1), 35–58.

Taylor, V. and Trentmann, F., 2011. Liquid Politics: Water and the Politics of Everyday Life in the Modern City. *Past & Present*, 211 (1), 199–241.

Turton, A. and Meissner, R., 2000. The Hydrosocial Contract and its Manifestation in Society: A South African Case Study. In A. Turton and R. Henwood, eds. *Hydropolitics and The Developing World*. CIPS/African Water Research Unit, University of Pretoria, 37–60.

van Holstein, E., 2015. Sharing Water: The Social and Technological Infrastructures of Resourceful Water Practices in Community Gardens. Royal Geographical Society with Institute of British Geographers, Annual Conference, Exeter, United Kingdom, 1–4 September.

Ward, S.L. and Butler, D., 2016. Rainwater Harvesting and Social Networks: Visualising Interactions for Niche Governance, Resilience and Sustainability. *Water*, 8 (11), 526–551.

Whatmore, S.J., 2013. Earthly Powers and Affective Environments: An Ontological Politics of Flood Risk. *Theory, Culture & Society*, 30 (7–8), 33–50.

Wild, T.C., Bernet, J.F., Westling, E.L. and Lerner, D.N., 2011. Deculverting: Reviewing the Evidence on the 'Daylighting' and Restoration of Culverted Rivers. *Water and Environment Journal*, 25 (3), 412–421.

Woelfle-Erskine, C., 2009. Emerging Cultural Waterscapes in California Cities Connect Rain to Taps and Drains to Gardens. In A. Lassiter, ed. *The Sustainable Water Reader: Lessons from California for the 21st Century*. Sacramento: University of California Press, 317–341.

Woelfle-Erskine, C., 2015. Rain Tanks, Springs, and Broken Pipes as Emerging Water Commons along Salmon Creek, CA, USA. *Acme*, 14 (3), 735–750.

Yates, L. and Evans, D., 2016. Dirtying Linen: Re-Evaluating the Sustainability of Domestic Laundry. *Environmental Policy and Governance*, 26 (2), 101–115.

Zifkos, G., 2015. Sustainability Everywhere: Problematising the 'Sustainable Festival' Phenomenon. *Tourism Planning and Development*, 12 (1), 6–19.

Chapter 13

Thinking like water moves

Living with climate change in Tarawa, Kiribati

Maria Louise Bønnelykke

Introduction: sinking islands

In the centre of the Pacific is a small island nation, Kiribati, one of the poorest nations in the Pacific faced with the detrimental effects of climate change. Sea level rise and saltwater intrusion into crops and freshwater aquifers are already commonplace. It is projected that by 2050 the islands will be completely overcome by the ocean or otherwise uninhabitable (Boncour and Burson 2010, 11).

This chapter is based on ethnographic fieldwork carried out in 2010–2011 exploring perceptions of environmental change among villagers and development practitioners on Kiribati's main Island Tarawa.[1] Here, the ocean eroded the coastline and seeped into freshwater resources, and according to recent reports from the family I lived with during fieldwork, extreme tides, known as king tides, now cover the entire island several times a year.

Faced with climate change the overarching debate concerning Kiribati revolves around issues of relocation – leaving the island. For example, the I-Kiribati citizen Ioane Teitiota made a claim to be recognised as a climate refugee in New Zealand, but was denied (Dastgheib 2015). And the official policy of the government is known as 'Migration with Dignity'. Within this framework, migration opportunities are negotiated with more resilient countries in the region for those who already want to migrate (McNamera 2015, 62).

On a local level, the villagers similarly reflected on what relocation meant for them. One respondent, Lysa, owned a video production company with her Australian husband. They produced short films to bring awareness about climate change on the outer islands. They started this project when their house was destroyed by a high tide coinciding with strong winds. She said to me in a manner that was not emotional, but simply matter-of-fact:

> When a child [that] is born here today [...] reaches the age of 16 there
> will be no land left to stand on, [even if] you have a bit of land you have
> lost your freshwater. If you are talking about leaving Kiribati, you are not

just talking about the person, but also the totems and the bones of their loved ones. So you are not just dealing with a living person, you are dealing with everything else.

While Lysa reflected on her home disappearing normal life carried on. When she finished talking she calmly stroked a cat that had jumped on her lap. The front door was ajar and outside a frangipani tree covered in white flowers leaned over the graves of Lysa's relatives. Occasionally, people passed by outside and called greetings through the door or a bus, playing loud pop music, disrupting the rhythmic sound of waves breaking on the reef in the distance.

Before lamentingly saying goodbye to these 'sinking islands' I will give a brief introduction to the impressive ancient formation of these atolls, and how people came to live in these harsh island environments. Through ethnographic material I will show how over centuries the emerging relationship between water, land, and humans can create novel thinking about the uncertain future facing people on the islands. I draw a comparison between the way the villagers have learned to live in changing environments, and the way water moves through landscapes, and I call this *thinking like water moves*. Perhaps to give some small hope I argue that understanding how the environment works and anticipating the future were still important to people on islands affected by global anthropogenic climate change.

The atoll emerging

Kiribati consists of 33 atolls and reef islands straddling the equator between 04°43′ N and 11°25′ S latitude and 169°32′ E and 150°14′ W longitude[2] (Thomas 2003, 2). Kiribati has a total land area of 726 km^2, but is spread across an ocean area of 3.5 million km^2 (Locke 2009, 173). The Gilbert Islands, a cluster of 16 islands in the Western part of Kiribati, dotting the equatorial Pacific in a line from north to south is where the majority of the approximately 100,000 inhabitants live. Some 1,400 km east of the Gilbert Islands you find the Phoenix Islands, consisting of eight largely uninhabited islands. Travelling further east you will find the Line Islands, also consisting of eight islands, among them Kirimati island, the biggest atoll in the world in terms of land mass (388 km^2), but hosting only 4 per cent of the population (Thomas 2009, 569).

Kiribati has a deep geological history – the slow and subtle formation of the atoll – and a more recent history of some thousand years with human settlement, social transformation, and climate change. Although the locals were not informed by geological studies there were some similarities between geological and local understandings of the atoll. Pacific studies explain that atolls are volcanoes appearing in belts across the Pacific Ocean. In some parts the Pacific Ocean floor is flat, but in the north and west, platforms and ridges rise above the water's surface to form what we today call Melanesia and Western Micronesia. In the South Pacific, Ron Crocombe describes how the islands of Eastern Micronesia, where Kiribati is located, are the tops of dormant volcanoes, which weigh heavily on the seabed causing it to sink a little (Crocombe 2001, 22). According to studies in marine biology this slow sinking has taken place over the past 3,000–4,000 years, known as the subduction of the Pacific tectonic plate under adjoining plates, creating processes of seismic activity in the area commonly referred to as the Ring of Fire. Because of this subduction the volcanoes have appeared to be sinking into the ocean throughout time. The atolls emerging from the ocean are the product of marine organisms, coral, which thrive in tropical waters. The coral, growing upwards, creates fringing and barrier reefs, which partly or completely separate an inner shallow lagoon from the outer ocean, creating the characteristic circular or semi-circular shape of the atoll (Grigg 1982, 29–30). Charles Darwin first conceived a theory of the atoll. He made the following entry in his diary, about the Tahiti islands, in 1836:

> Hence if we imagine such an island after long successive intervals, to subside a few feet in a manner similar but with a movement opposite to the continent of South America; the coral would be continued upwards, rising from the foundation of the encircling reef. In time, the central land would sink beneath the level of the sea and disappear but the coral would have completed its circular wall. Should we not then have a coral island? Under this view we must look at a lagoon island as a monument raised by myriads of tiny architects to mark the spot where a former land lies buried in the depths of the ocean.[3]
>
> (Darwin in Grigg 1982, 29)

The sandy atolls are at once persistent, land is slowly but steadily forming in the world's largest ocean, and elusive, as they are dynamic in their response to seawater erosion and deposition. The legend of Makin Island in the north of Kiribati is an example. Villagers told an expedition team from the United States of a vanished island about two days' sail to the northeast of Makin known as Tarawa ni Makin (Hale in Thomas 2009, 578). Similarly, oceanic geoscientist Patrick Nunn, when exploring vanished islands in the Pacific Ocean, distinguishes between catch-up reefs, reefs able to grow upwards as postglacial sea levels rose, and give-up reefs where the coral drowned as the seawater rose (Nunn 2008, 38–39), and in contemporary Kiribati houses and sometimes entire villages are moved further inland as the ocean erodes the islands. Taking a glance at the deep geological time suggests that Kiribati has been in the process of becoming for thousands of years. It did not appear as a place when the islands were stable enough to support human settlement (Dickinson in Thomas 2009, 576), when a Spanish boat searching for Terra Australis Incognita first visited the islands in 1537, or when Europeans had sighted all the islands of Kiribati in 1820 (MacDonald 1982, 14). The islands have a long history of their own.

How and why these remote atolls became settled has been passed on in the local oral history and migration stories, but within a scientific understanding the issue of settlement has caused debate. Kiribati as a remote and fragmented island landscape seems to have informed earlier hypotheses, which suggest the islands were settled by people who had been driven to the ocean by hunger, and were travelling on voyaging canoes drifting in the Pacific Ocean. About the Gilbert Islands, furthest to the west, geographer Woodford writes: 'At the time of their first inhabitants, drifting, no doubt, from their former habitation, by accident going they knew not whither, and seeking a land they knew not what' (Woodford 1985, 329). On the contrary, contemporary hypotheses suggest that the Gilbert Islands have experienced several waves of migrant settlers, and that these journeys were carefully planned rather than accidental. Archaeological studies suggest that the first settlers came from the west, from Melanesia, perhaps some 4,000–5,000 years ago. About 3,500 years ago there was another movement from southern Melanesia to the Fiji-Tonga region. Such began the Polynesian society and culture, which has been carried as far as New Zealand and Hawai'i. Historical and ethnographic literature describes Gilbertese myths telling similar stories of migration from Samoa to the Gilbert Islands, and after many generations a wave of return migration along the same trails (Grimble 1970, 151; MacDonald 1982, 1–3; Thomas 2003, 4–5; Uriam 1995, 22). This more recent hypothesis rejects the idea of the settlements as accidental, arguing instead that these migration journeys required skilled navigators and careful preparation. The two hypotheses suggest not only how the islands were settled, but also a perception of the islanders settling them, as either vulnerable, drifting, and finding land by accident or as resilient and skilled navigators.

Archaeological research shows that as the islands became settled, humans began making their imprint on the environment, introducing new species and agricultural techniques. The oldest traces of human occupation tended to be on wider islands with sufficient freshwater to support agricultural production. This was usually where the babai, commonly known as *taro*, pits were dug. Soil was excavated in order to cultivate the babai straight into the surface of the groundwater – a fresh body of water that rests on top of the denser surrounding saltwater (Thomas 2009, 582). R.L.A. Catala's studies of human ecology in Kiribati state that coconut trees today cover large areas of the islands, partly as forest and partly planted by landowners (Catala 1957, 21). The breadfruit trees (*te mai*) provide shade and its fruits are starchy and filling and served with meals much like potatoes. The pandanus tree grows large green and orange fruits with a spiky appearance. The fibre-rich fruit is eaten raw or prepared with coconut cream, and the leaves of the tree are used for roofing, wrapping, and rolling tobacco into long slim cigars (ibid., 55–56, 58). Some species of coconut, breadfruit and pandanus trees are considered indigenous and others are introduced. The babai is an introduced species (Thomas 2003, 12).

Human settlement thus required modification of the landscape, for example through the introduction of new species. Without it, it would have been difficult for anyone to survive on these islands (Thomas 2009, 582). However, studies point out that the modification has now been overtaken by degradation, in particular on South Tarawa. A geographical study by D. Storey and S. Hunter identify that degradation is mainly caused by overcrowding, squatter settlements on water reserves, lagoon pollution, and poor solid waste management and sanitation services (Storey & Hunter 2010, 168). The population living on South Tarawa continue to cultivate the local crops described above, but they have become entirely dependent on imported foods to meet basic dietary needs, and the plastic packaging adds pressure on solid waste management (ibid., 173).

Sweeping through the history of Kiribati as I have done above allows me to detach myself from the situatedness of my fieldwork, and momentarily explore the history of engagement between people and the environment on the atoll. It is intended to demonstrate that much of what happens today is conditioned in varying degrees by what happened yesterday, and much of what happens tomorrow is influenced by forces already in operation. And despite this continuity, nothing is determined; the islands continue to be simultaneously persistent and elusive, like Tarawa ni Makin, which was inhabited until it vanished under the ocean. History supports both consistency and change, certainty and uncertainty, and the islands are shaped in connections between water, land, and human practices. One informant who had studied marine biology at the University of the South Pacific explained that sometimes, in the southern islands, they still experienced earthquakes from the volcanic activity:

> We do have earthquakes here in most of the Southern Islands. It is the volcanoes; they are still active in the south. That is also why the three most southern islands have no lagoon, and they are higher above the sea level. They are not finished forming yet. But I am very careful [who I suggest this to] because in the creational stories every island wants to be the first. In Nikunau they have a God, Riiki, who created the island and he also created the ponds. But I think that those ponds are the beginning of the lagoon. Maybe in three generations from now, we will see the ponds will actually be the lagoon. But I am very careful [about suggesting] this to the old men because I don't want to contradict their stories.

The creational stories conveyed how ancient gods shaped the islands and the ponds at the beginning of time, while science related earthquakes to active volcanoes under the sea and ponds to an emerging lagoon. Both ideas demonstrate the stability and change, certainty and uncertainty the islands have undergone, and that meaning-making carefully considered how materialities interact, move, and change. In this way, the opposition of a stable and certain world against a changing and uncertain world are not mutually exclusive. Rather, such dualities exist in an interconnected and organic relationship. When observing how water carves out a landscape and the way it moves and flows, both sustaining and destroying, we learn that dualities are not mutually exclusive opposites. In fact, dualities, such as certainty and uncertainty, consistency and change, are identical in nature, but different in degree.

Knowing water, knowing landscapes

Knowing water was not just technical or scientific, water was part of the villagers' existential framework, and it played an important role in land tenure and kinship. While I carried out fieldwork I also engaged in knowing water. I had to know 'what the tide was doing'. During high tide I had to wade through seawater, observing small fish scattering under my feet, to make the short walk from the road to my house. I also had to learn how to get clean drinking water, and how to keep it clean. Gradually, I learned the routines that were necessary to live with and know water. Once I returned home from fieldwork, the documentary *The Hungry Tide: A Personal Story About a Pacific Nation on the Front Line of Climate Change* (2011) was released about an I-Kiribati woman, Maria Timon, and her work advocating for Pacific Islands. In the opening scene a woman and her son waded through the knee-high water. Next the documentary showed a man and his struggle to keep the tide out of his house by digging channels. His efforts to control the ocean failed, and as a last resort, but without any sign of panic, he ordered his family to grab their belongings that were floating away with the sea. Sitting in the comfort of my home in Denmark, where the ocean never reached my doorstep, I suddenly wondered how people live like this. A question that in a paradoxical way was never that pertinent during my fieldwork because there, people had to live like that. After returning home water was and was not the same, simultaneously. Mol and Law (1994, 641) develop the concept of a fluid where an object is simultaneously similar and dissimilar. Looking at how anaemia is diagnosed differently in Zimbabwe and the Netherlands they argue that anaemia is a fluid, which means that there is no ontological rupture in the diagnosis of anaemia in Zimbabwe and the Netherlands, but they are not the same either (ibid., 658). It is not one single clinical network with elements that hang together that transports anaemia from Zimbabwe to the Netherlands. Neither is it two regions each with their own methods to diagnose anaemia. Rather anaemia is a fluid, which 'generates the possibility of invariant transformation' (ibid.). The authors are concerned with the question: how can something (anaemia) that is so different be the same? When I returned home from fieldwork I questioned how the same thing (water) can change so much? Water was a fluid in more than the obvious way. It was the same fluid substance, but it multiplied through practice and became notoriously difficult to define. De Laet and Mol (2000) later develop the notion of fluidity to describe a water pump known as the Zimbabwean bush pump type B, and they define this pump as a fluid technology. It is a mass-produced water pump, but its components and boundaries are intentionally vague and makeshift. The strength of this vagueness is that the pump travels easily between different contexts and locations without imposing itself or being too rigorously bounded or situated. It is adaptable, flexible, and responsive (ibid., 225, 233). Water is, like the bush pump, a single multiple object (Mol 2002, 142).

With the onset of climate change in Kiribati, the local landscape also became increasingly fluid – it simultaneously changed and remained the same. When Amon, a man in his fifties, sailed me across the lagoon one morning, he explained how fish stocks and fishing practices had changed during his lifetime:

> We used to go fishing in October and November for a fish called te manoku. We don't see those fish anymore, not at all. It is the westerly winds that bring that fish, but now because there is no wind, the fish can't come. The red snapper has also changed. They are not in the places where we used to find them, but now we have discovered new fishing grounds where we can fish them. In the 1960s and 1970s there were few sea walls, but because of erosion people started building. After people started building the sea walls the coral started growing very fast on the ocean-side of our land. We saw no coral here before. In the 1990s we suddenly saw the coral growing from the break towards the land. They were not there before; it was flat stone before. In the past, it was raining three or four times a week on Butaritari, the island where I grew up. Now it rains maybe once a week. When the drought is severe the top of the coconut trees falls off. Then we know that salinity in the ocean is high, and that it will be a good season for the octopus.

In his explanation, Amon was sensitive both to how the environment changed in unpredictable ways – the westerly winds failing to blow – and to how human practices shaped the environment, such as the construction of sea walls making the coral grow. Amon did not try to rule out uncertainties; he accepted that he was living in a fluid environment. There are several examples of how the uncertain and changing conditions of the island environment have shaped practices since human settlement. In precolonial times families were restricted in size to manage the changeable resources available – local medicine was used to induce abortion, infanticide was performed, adoption to childless families was practiced, and furthermore foods were carefully prepared and stored for times of drought (MacDonald

1982, 13). A local marine navigator also explained that Bakoa, the shark and the spirit of the ocean, and Tabwakea, the sea turtle and the spirit of the land, have since the beginning of time been in a competitive and struggling relationship. However, ways to reconcile their struggle did exist:

Tabouea:	Sometimes when you want to go to the sea, you bring Tabwakea, and when you come out from the sea you bring Bakoa.
Maria Louise:	What do you mean?
Tabouea:	I mean all the natures of Bakoa, you bring it to Tabwakea, just to try and make them friendly with each other.
Maria Louise:	What can you bring from the ocean for example?
Tabouea:	Some of the grass from the deep ocean, like trees [seaweed] from the deep ocean. You bring it onto land. And you bring the flowers from the land and give them to the ocean.

You sense the uncertain relationship between ocean and land in this ritual. The small sandy islands, and the villagers' dependency on the resources of the ocean. However, when faced with uncertainty there was no rupture in the meaning, instead meaning moved and flowed. In other words, the environment did not come undone because of erosion or the disappearance of te manoku in Amon's story. When faced with uncertainty, like the onset of prolonged drought, new connections were crafted, people started catching octopus.

Thinking like water moves

In his book, *After Method: Mess in Social Science Research*, John Law refers to the social world as unspecific, slippery, ephemeral, and changing like a kaleidoscope or not having much of a pattern at all. Law asks how social science can begin to know a world that is so unpredictable without distorting it into clarity, and whether knowing is even the right metaphor (Law 2004, 2).

It is not only social scientists who are preoccupied with knowing the world, as the empirical material in this chapters shows, so are other people who occupy it. I want to emphasise two characteristics of moving water to use it as a comparison to how the villagers thought.[4] First, water is in fact never stable, it is always in flow, and its form always tentative. Second, there is no contradiction between water's ability to sustain and destroy, it is always doing both. Water connects dualities and thereby shows that what seems to be in opposition is in fact identical in nature, but different in degree.

The villagers did not attempt to pin down the uncertainty of the environment by distorting it into clarity. To know did not require a stable or non-shifting environment, rather the villagers' thinking could be compared to the way water moves, always in flow, unstable, and in search of new patterns as old ones break down. The marine scientist quoted earlier in this chapter flowed between the scientific explanation of how the islands were shaped in myriad connections between sand, coral, ocean, and volcanoes, and the traditional stories of creation. Whereas Amon was constantly attuned to disappearing environmental patterns and the emergence of new patterns when he described how sea walls can make new coral grow or increase erosion on adjoining land, and how he observed connections between salinity and octopi, westerly winds and *te manoku*.

When thinking like water moves the subject comes to matter; the subjects were the meaning makers. They were sensitive to stability and change, to certainty and uncertainty. Meaning was in constant flux, but that did not make it collapse or entropic. As connections came undone new ones were crafted and patterns emerged. When the villagers thought like water moved, making the world meaningful did not require a stable and certain world. Instead uncertainty and partiality characterised the villagers' thinking – like water meaning flowed and moved. When the ocean broke down a nearby sea wall separating ocean from land and protecting his house Paul, my neighbour, succinctly stated: 'The water *will* find a way', thereby in one sentence stating that the world is both certain and uncertain. Another example was Taati, who I lived with, and who one day gazed over a large tidal flat where a family had built a home that was flooded by the tides almost daily. I asked her: 'What do you think? Do you think this is climate change?' Taati replied:

A lot of journalists have filmed this place as a place that has been affected by climate change. But I don't know. I always make sure to tell people that this could be climate change, but I also think that the water wants to flow here. Maybe it is people living where they are not supposed to live.

People thought like water moved, and water moved peoples' thoughts. Water taught Paul and Taati how to think, and they engaged in connecting environments through water. They were sensitive to the (un)certain movement of water, and it shaped the way they thought. These are examples of *thinking like water moves*, where human ideas about how the material world is connected make the environment emerge. Another example could be the way coastal engineers, when working with hydrodynamic modelling, tried to manage the intensely complex and unpredictable process of waves approaching a reef system – no two waves will ever be identical, they explained to me, and therefore they were aware that their coastal models reduced a reality that was too complex to know. Thus, their models and understanding of the environment were, like water, partial, unfinished, and on the move.

Thinking like water moves allowed the villagers to know the world in its tentative form and to connect contradictions. In other words, it allowed the villagers to respond to and live in a changing world. As described earlier there were two hypotheses about how the islands of Kiribati were settled. One suggested the islanders were drifting in the ocean and only by luck did they come across land. The other recognises the islanders as skilled navigators who, well prepared, set out to sea in search for new land. By describing how the villagers come to know the world around them by thinking like water moves we are reminded that they are not just hopeless victims lost in a changing world. More accurately these villagers are responding to and in the process of knowing their world.

Notes

1 The empirical material I draw on in this chapter comes from semi-structured interviews and participant observation among key informants, both villagers living on Tarawa and development practitioners working there.
2 I write these geographic coordinates not necessarily expecting the reader to thereby know the location of Kiribati, but to underline the fluid environment in which Kiribati is located, and the way these coordinates help identify locations in ocean environments.
3 New geological studies show that it is not only coral growth and volcanic subsidence that shape atolls. The different morphology in atolls, from accreting fringing and barrier reefs to reef terraces, has been shown to be a result of a combination of coral growth and volcanic subsidence together with glacial sea level cycles (see Toomey et al. 2013).
4 Other scholars have explored how people think like water. Gaston Bachelard, philosopher in the field of philosophy of science, has coined the concept of water mindset, which enables people to participate in the elemental material of water (Bachelard 1983, 5). Bachelard explains that water invites people to explore their aquatic reflection, which, unlike a mirror, provides depth and continuity; an image the superficial reflection of a mirror cannot provide. Self-reflection, and not move-ment, is central to Bachelard's water mindset. He refers to Narcissus, who loves his own reflection in the water because what he sees is not superficial but through his watery reflection he comprehends his own embodied thickness (ibid., 19–43).

Bibliography

Bachelard, G. (1983) *Water and Dreams: An Essay on the Imagination of Water*. Dallas: The Pegasus Foundation.
Boncour, P. and Burson, B. (2010) Climate Change and Migration in the South Pacific Region: Policy Perspectives. In B. Burson (ed.), *Climate Change and Migration: South Pacific Perspectives*. Wellington: Milne Print, pp. 5–28.
Catala, R.L.A. (1957) Report on the Gilbert Islands: Some Aspects of Human Ecology. *Atoll Research Bulletin*, 59, 1–187.
Crocombe, R. (2001) Place: Environmental Deterioration and Enhancement. In *The South Pacific*. Suva, Fiji: University of the South Pacific, pp. 22–44.
Dastgheib, S. (2015) Kiribati Climate Change Refugee Told He Must Leave New Zealand. *Guardian*. www.theguardian.com/environment/2015/sep/22/kiribati-

climate-change-refugee-told-he-must-leave-new-zealand (Accessed September 22, 2015).

De Laet, M. and Mol, A. (2000) The Zimbabwe Bush Pump: Mechanics of a Fluid Technology. *Social Studies of Science*, 30(2), 225–263.

Grigg, R.W. (1982) Darwin Point: A Threshold for Atoll Formation. *Coral Reefs*, 1(1), 29–34.

Grimble, A. (1970) *A Pattern of Islands*. London: John Murray.

Law, J. (2004) *After Method: Mess in Social Science Research*. London: Routledge.

Locke, J.T. (2009) Climate Change-Induced Migration in the Pacific Region: Sudden Crisis and Long-Term Development. *Geographical Journal*, 175(3), 171–180.

MacDonald, B. (1982) *Cinderellas of the Empire*. Canberra: Australian National University Press.

McNamera, K.E. (2015) Cross-Border Migration with Dignity in Kiribati. *Forced Migration Review*, 49, 62.

Mol. A. (2002) *The Body Multiple: Ontology in Medical Practice*. Durham, NC: Duke University Press.

Mol, A. and Law, J. (1994) Regions, Networks and Fluids: Anaemia and Social Topology. *Social Studies of Science*, 24(4), 641–671.

Nunn, P. (2008) Islands that Vanished Long Ago. In *Vanished Islands and Hidden Continents of the Pacific*. Honolulu: University of Hawai'I Press.

Storey, D. and Hunter, S. (2010) Kiribati: An Environmental 'Perfect Storm'. *Australian Geographer*, 41(2), 167–181.

Thomas, F.R. (2003) Kiribati: 'Some Aspects of Human Ecology', Forty Years Later. In *Atoll Research Bulletin*, No. 501, Washington, DC: Issues by the National Museum of Natural History, Smithsonian Institution, pp. 1–40.

Thomas, F.R. (2009) Historical Ecology in Kiribati: Linking Past with Present. *Pacific Science*, 63(4), 576–600.

Toomey, M., Ashton, A. D., and Perron, J.T. (2013) Profiles of Ocean Island Coral Reefs Controlled by Sea Level History and Carbon Accumulation Rates. *Geology*, 41 (7), 731–734.

Uriam, K.K. (1995) *In Their Own Words: History and Society in Gilbertese Oral Tradition*. Canberra: ANU.

Woodford, C.M. (1985) The Gilbert Islands. *Geographical Journal*, 6(4), 325–350.

Chapter 14

Narratives that travel

Anxiety, affect, and water politics in the Deschutes watershed of Central Oregon

Kirsten Rudestam

Introduction

The Deschutes Basin, a watershed spanning central Oregon, is one of countless regions across the American West experiencing an increasing demand for water amid a rapidly decreasing supply. The human population in the Deschutes has the fastest growth rate of any county in Oregon, but while municipal demand has skyrocketed, available surface water supplies are already over-allocated. In addition, during the spring and summer irrigation season, water diversions cause a dramatic reduction in the Deschutes River's flow, contributing to degraded fish habitat and poor water quality.

The Deschutes is emblematic not only for its water supply concerns. The basin is nationally renowned for having undertaken an innovative approach to solving its water distribution problems. In 2001, tribal members, irrigation district managers, and environmental proponents came together and established an institution that uses collaborative mechanisms for managing and distributing the basin's freshwater supply. Since the inception of the Deschutes River Conservancy (DRC), collaborative management of water has become increasingly widespread across the American West and the Deschutes has served as a role model for many of these initiatives.[1]

How and why did this new strategy for managing shared waters take hold? In this chapter, I examine one important event that helped catalyze the collaborative approach to water management that emerged in the Deschutes – the conflict over water in the Klamath Basin just south of the Deschutes, where a year of drought incited intense conflict between irrigators, tribal members, and environmentalists. While the case was extraordinarily complex, it became neatly summarized as a "fish versus farmers" scenario, and non-tribal Deschutes water users became anxious that a similar situation would play out in their own basin. This fear was foundational in motivating the innovative participatory water strategies that were subsequently adopted in the Deschutes and that have since been emulated by other western basins.

In drawing a comparison between the neighboring watersheds, I make two theoretical interventions. For one, I highlight world-making as an intrinsically relational process. Second, I demonstrate that movements in and changes of policies over time can be mobilized through and via affect. In this regard, I argue that acknowledging the political force of feelings is central to fully understanding the adoption and rejection of contemporary water policies.

Theorizing affect and water governance

In describing the force of feelings, I am deliberate in using the term "affect," and my decision to do so requires further elaboration. A concept and theory that has become increasingly compelling to cultural theorists, affect is taken up in different ways by different scholars, and its diversity of interpretations can often result in theoretical vagueness and confusion. I understand affect to be that which encompasses the breadth of public feelings, material and sensate experiences and perceptions that may have not yet been linguistically or conceptually captured (Massumi 2002). I draw largely from Brian Massumi (2002, 2015) who situates affect within a lineage of process philosophers such as Spinoza, Henri Bergson, Felix Guattari, and Gilles Deleuze.

Spinoza described affect in deceptively simple terms as the power to affect and to be affected. What this entails is both contact and receptivity or, in Massumi's words, "to be open to the world, to be active in it and to be patient for its return activity" (2015, ix). What differentiates affect studies from other process-based ontologies is its emphasis on change via the intensities of feeling and emotion that invariably accompany encounters between subjects and their subsequent transformation. Affect includes awareness, conscious thought, and cognition, but it also foregrounds embeddedness and embodiment, and the ways in which the body senses change.

While affect scholars in the philosophical tradition have been deliberate in differentiating affect as independent from emotion, others have understood emotions to be centrally related to and at times inseparable from affect (Ahmed 2004; Berlant 2011; Williams 1978). In marking moments of transition, affect accompanies (perhaps even defines) every encounter, and the feeling of change, or how it registers in the body, is often expressed via emotional states. Sarah Ahmed (2004) describes emotion as a contracted or mediated form of affect that works to shape the "surfaces" of individual and collective bodies. Her theory of affective economies explains how the circulation and accumulation of particular affective states and emotions produces subject positions to which people feel they belong. Likewise, cultural materialist Raymond Williams (1978) describes emotion as central to social and political contexts. His idiom "structures of feeling" refers to the affective, inchoate forces that exert pressure on present-day experience and whose emergent properties are most recognizable in cultural forms, such as art and literature.

I incorporate these perspectives on affect, emotion, and the political to better understand how and why collaborative water management strategies were adopted in the Deschutes Basin. Conventional approaches to studying natural resource management often utilize concepts such as policy transfer and policy diffusion to explain the travel and uptake of various policies over time (Huitema & Meijerink 2017). In critiquing such approaches, political scientist Mukhtarov (2014, 71) argues that these theories tend to "assume the perfect rationality of actors, the stability of governance scales, and the immutability of policy ideas in their travel." He rejects such presumptions that policy uptake is based on rationality and linearity and suggests that we instead recognize policy ideas and problems as fluid in time and space. While Mukhtarov's alternative ("policy translation") does, like affect, place attention on process rather than outcome, I suggest that applying theories of affect to our under- standings of policy uptake provides us with an important dimension missing from his account. Affect theory helps us to recognize the importance of bodily encounters and collective affective conditions in the context of how and why policies move.[2]

Methodology

In the natural resource management literature, the Deschutes Basin has a winning reputation for managing to maintain irrigators' senior rights to water while plumping up instream flows. But how and why did the collaborative strategy in the Deschutes take hold? And whose and what interests does it serve? In answering these questions, I engaged in an ethnography of water politics, of human–water encounters, and of water itself as it moves and is moved through the Deschutes waterscape. Over the course of four years, I conducted 45 formal interviews and approximately 30 informal interviews with environmental activists, landowners, farmers, irrigation district operators, tribal members, and other community members. I found participants primar- ily through local networks and snowball sampling, with a main intent of including a wide range of perspectives from self-identified "stakeholders" in the local water supply (Schutt 2009). In addition, I worked for a total of six months as a participant intern with the DRC, spent time informally with residents, attended community meetings and attended over 50 professional meetings with water managers. I also engaged in a discourse analysis of representations of water issues and regional identity in the Deschutes Basin through historical documents and local and national media. I transcribed all interviews and field notes, and coded these and all archival files using grounded theory.

My time in the Deschutes revealed that while collaborative governance is becoming characteristic of contemporary water management practices, the empirical and conceptual focus on these new management schemes overlooks the ways in which feelings and affective encounters shape the uptake of political practices and the ways in which water moves and is moved through the landscape. In making sense of how and why the Deschutes adopted the management strategies that have since hallmarked its "success," I found that the events that unfolded in the Klamath Basin and their affective associations were crucial to the evolution of water policy in the Deschutes.

Before turning to an analysis of how affect and emotion mobilized water policy in the Deschutes, I begin by situating residents' water experiences in the broader processes of historical change. The following section illustrates the social and environmental transformations that were a product of colonial and capitalist expansion in the Klamath region. I describe how the water conflicts that ensued in the wake of these developments were narrated with a simple normative polemic – "fish versus farmers." I then demonstrate how the affective weight of this narrative and of the Klamath incident in particular translated across the geographic border, motivating alternative water management practices in the Deschutes.

Water wars in the Klamath

The Klamath Basin, situated just south of the Deschutes in Southern Oregon, was originally occupied by the Klamath, Modoc, and Yahooskin Band of Snake tribes (called "the Klamath Tribes") who extensively utilized the local waterways for food, trade, and travel. Akin to events that played out in the Deschutes, early colonial efforts to irrigate the land transformed the Klamath landscape and the lifeways of native peoples. By the twentieth century, the Klamath Tribes had lost their tribal status as well as nearly all of their traditional lands (Doremus & Tarlock 2008) and the Bureau of Reclamation had severed anadromous fish passage and transformed the basin into a highly maintained network of irrigation canals, shuttling water to various agricultural entities and ranchers (BOR 2009).

In 1975, water claimants in the Klamath Basin began a lengthy and conflict-ridden adjudication process, precipitated by Klamath tribal members' desire to clarify their water rights. Water adjudication involves assigning water rights to claimants based on their priority date, and in the case of the Klamath, over 700 people and institutions, including the Bureau of Indian Affairs, the US Forest Service, and other governmental agencies, made a case for their senior rights to local waters. Over 5,600 existing water users flooded the courts to oppose the adjudication, fearing that their water rights would be jeopardized by tribal recognition. Their fears were warranted; the lawsuits eventually determined that the Klamath tribes were, in fact, owners of the most senior water rights in the basin (Doremus & Tarlock 2003).

The adjudication process was just the beginning of what became a long and conflictive legal battle over local waters. In 1988 the short-nosed sucker and Lost River sucker were listed as endangered species under the ESA and in 1997 the coho salmon followed suit. The designation of endangered species complicated an already contentious relationship among existing water users, some of whom had felt unjustly robbed of their colonial water rights in the recent adjudication. In 2001, these users (primarily irrigators) were forced to make even more cutbacks to their water use. The US Fish and Wildlife Service (USFWS) and National Marine Fisheries Service (NMFS) demanded that irrigation water be modified to provide for the listed species' critical habitat. The coho needed more water released below the dam, and the sucker fish required more water left in the lakes above the dams. The water reclamation project interfered with both of these recommendations, and subsequently the US District Court ordered that all irrigation be halted.

The federally mandated halt of water resulted in uproar. Thousands of upset irrigators and sympathetic citizens took to the streets in a passionate demonstration of anger and resistance, parading in a "bucket brigade" and protesting in front of the government center in Klamath Falls. Activists even illegally breached the headgates of the dam. Overnight, the sleepy town of Klamath Falls erupted into a maelstrom of protest. Even the Bush administration joined the fray, sympathizing with the irrigators and commissioning a new study released by the National Research Council (NRC) that refuted the biological opinions set forth by the NMFS and the USFWS. Based on the NRC study, the Bureau of Reclamation (BOR) created a new management plan that authorized water use, and Secretary of State Gale Norton flew to Klamath Falls to ceremoniously open the dam's headgates.

For the next few years, the BOR operated on an annual basis, and continued to provide water to irrigators despite the listing of ESA species in the region. But in 2002, thanks to a drought year, thousands of coho and chinook salmon died in their seasonal migration to the ocean, and a slew of conservation groups filed a lawsuit that led to the rejection of the BOR's operations. The standoff between Klamath farmers and the federal government softened with the dawning recognition of a potential bigger threat to colonial irrigators – the Klamath tribes – who had managed to secure the most senior water rights in the basin in the adjudication process. In 2005, talks commenced between irrigators, government officials, environmental groups, and tribal members, to come to an agreement around water use. The resulting Klamath Basin Restoration Agreement (KBRA) took years to produce, and remains a contentious arrangement (Doremus & Tarlock 2008).

Narratives that travel: fish versus farmers

Despite the complexity of the case, the publicity around the Klamath Basin conflict relied on a trope of "fish versus farmers." Signs held by protesters at the time read, "Call 911, some sucker stole our water" (in reference to the endangered sucker fish). A headline from an article from SF Gate declared, "Fish Versus Farmers in Conflict Over Klamath River: Spawning Fish Vie with Farmers in Dispute Over Klamath Waters" (Fimrite 2013). Doremus and Tarlock (2003) wrote what is perhaps the most frequently cited academic article on the Klamath case entitled, "Fish, Farms, and the Clash of Cultures in the Klamath Basin." Although the title of their piece indicates that there is more to the story than fish and farms, the public has overwhelmingly characterized the event as a crisis centering around these two entities, writing out the conflict between Klamath tribes and farmers and the ways in which the deliberations ultimately (although anemically) enhanced indigenous rights.[3]

Fish versus farmers is not a new polemic. The "jobs versus environment" mantra has played out in various forms throughout rural America. Rebecca Scott, in researching the Appalachian coal industry, describes the trope as hegemonic, "reflect[ing] a well-worn articulation between a particular conception of the human relationship to nature and a notion of nationalistic progress" (Scott, 2010). The "jobs" referenced by the phrase are almost always natural resource extraction or heavy industry centered, seen to be the backbone of the economy, whereas the "environment" evokes images of national parks and spotted owls.

In the case of water, we see jobs versus the environment playing out in the pitting of "fish" against "farmers." For example, an *Associated Press* article described a local water controversy as a case wherein "The federal government shut off water to most of the farms in 2001 to protect the salmon" (Barnard 2013). In 2014 House Speaker John Boehner supported a bill to roll back environmental protections of the California Delta, claiming, "How you can favor fish over people is something people in my part of the world would never understand" (Goodyear 2014). Likewise, a *Washington Post* headline from 2009 read, "It's Farmers Versus Fish for California Water" (Richardson 2009), in response to mandatory water cutbacks initiated by the Environmental Protection Agency to protect the endangered delta smelt.

The translation of complex issues into a fish versus farmer polemic has a number of consequences. For one, it can lead to the avoidance of more pertinent and pernicious issues underlying current water conflicts. While environmental protections exert pressure on farmers to cut back their water use, a number of other factors impact the quantity of water available for irrigation, such as climate change and inefficient irrigation infrastructure (Bacher 2009; Miller 2014; Orr 2014; Overstreet 2014). But in blaming environmental protection of endangered species for water deprivation, politicians and farmers routinely ignore the multifaceted factors that influence water availability.

In addition, the fish versus farmers polemic perpetuates the image of irrigators as family farmers with individual, hard-won water rights when in reality this kind of landscape and livelihood is increasingly rare. In the Deschutes and in the Klamath, the arid climate, poor soil conditions, and swelling urban and suburban populations have hindered the success of agricultural operations, impacting ranchers and farmers long before the emergence of federally listed species. In many ways, fish versus farmers speaks to a larger antagonism that has become increasingly apparent between white rural residents of this country and government interventions (Hochschild 2016).

Despite the shortcomings and oversights inherent in the fish versus farmers frame, it has become a primary narrative shaping water politics of the American West. In addition to its visibility in mainstream media and historical accounts, almost all of the irrigators and ranchers that I spoke with during my time in the Deschutes reduced contemporary water issues into competitions between farmers and fish. If we seek a more sophisticated understanding of complex water ecologies and how, when, and why particular water practices are enacted, we cannot ignore the affective grip of this particular narrative, or the capacity of this narrative to travel.

Differences that matter

As neighboring basins, the Deschutes and the Klamath share many of the same characteristics. In both, Euro-American colonists displaced native people and transformed the local waterways to accommodate irrigation needs. Up until the late 1900s, both had economies that relied primarily on resource extraction, and in both the pressures from federal endangered species legislation and from Native American's insisting on their right to water interfered with colonial irrigation practices.

But the two basins contain important differences. The economy of the Deschutes has been bolstered by tourism and recreation, and increasingly ex-urban migrants have been relocating to the region, bringing with them new values for local landscapes and new forms of income generation. The Klamath has not witnessed the same demographic shift; its local economy continues to rely on resource extraction. Geographically the areas also differ; the Klamath River is flashy, responding to drought conditions by immediately dropping its levels, whereas the Deschutes has one of the most stable flows of any river in the western United States. Finally, the relative power of tribal interests in the two regions is also significantly different; the Warm Springs Tribal Nation in the Deschutes managed to maintain access to some of the most economically valuable waters in the west, while the Klamath lost not only their land but their tribal recognition.

Despite the differences between the two basins, what is clear is that in order to understand how and why the innovative water practices in the Deschutes took hold, we need to consider how the narrative of fish versus farmers traveled from Klamath northward. Farmers, environmentalists, and politicians in the Deschutes invariably referred to the Klamath as a case that motivated the unique water practices in their home basin. Conversations with various staff members and board members of the DRC indicated that the ability of the organization to secure funding and encourage collaboration was due in large part to witnessing the events that unfolded in the Klamath. For example, the DRC's public relations director said: "We're focused on a water management strategy so that in 50 years we're not a Klamath Basin."

Likewise, Davie, DRC board member, Warm Springs tribal member and director of Natural Resources, said,

> We have one of the best collaborative groups in the state [the DRC] ...
> Everyone is coming together and identifying challenges before they arrive
> ... That's the group that says we don't want to be the next Klamath. We
> don't want that situation to play out.

It is perhaps ironic that the fear of becoming "the next Klamath" became so prolific given that the likelihood of a similar situation occurring in the Deschutes was slim. Those working in local water politics agreed that while the Klamath served a role in motivating new water policies in the Deschutes, a parallel story could never unfold in a basin so fundamentally different from its neighbor. Lisa, from the DRC, spoke to this:

> You've probably encountered this, and I don't know if it's true or not ...
> Everyone points to the Klamath and says, "we don't want that to happen
> here." And folks in the Deschutes say that's not going to happen here
> because we're organized and have consensus groups and institutional
> ways that we work through these problems ... My own view is that
> you have very different populations you're dealing with, and some things
> are destiny and demographics are destiny, and issues are different and
> geography is different and ways issues are pressing down on you are
> different ... In some ways, I don't think push has come to shove in the
> Deschutes the way it has in the Klamath.

Lisa recognized that the threat of becoming the next Klamath was a significant motivator in the Deschutes, but acknowledged the low likelihood of it actually happening. A representative from Oregon Department of Water Resources who works in both the Klamath and the Deschutes shared a similar perspective: "It's not really a fair comparison [between the Klamath and the Deschutes]. The Klamath has multiple tribes, multiple species, the federal government is invasive, they have refuges, layers and layers of restrictions. Compared to that, the Deschutes is easy." Likewise, a local resident and member of the DRC board articulated a more textured understanding of the Klamath, and pointed to the complexity of the case:

> In the Klamath there was an ESA issue, but there was more than that.
> It was a cultural issue between irrigators and tribes, and it was a
> national issue as well that played out in national politics between
> Democrats and Republicans ... There it took a crisis for those parties
> to sit down and try to solve the problem. Here we don't have a crisis
> of that magnitude yet ... I think the Klamath is the closest example,

it's the closest to home. But every water problem I think is going to have its unique features.

This resident spoke to the unique conditions inherent in the two distinct regions, and offered an insightful observation about how, in addition to making waves in the Deschutes, the event incited partisan action at the national scale. Although the Endangered Species Act may pose a threat in both basins, there are and were a number of characteristics of the Klamath that simply do not exist in the Deschutes. Gil, fish biologist with Oregon's Fish and Wildlife Service, spoke to this disparity as well, insisting that "The Klamath is just worlds apart from the Deschutes in terms of biology and legal exposure, absolutely apples and oranges."

"We don't want to be the next Klamath!"

Despite the perhaps unrealistic potential of the Deschutes turning into "the next Klamath," people were clearly motivated by the events that had unfolded there, and they were motivated not so much by the political legislation or the real ability of a crisis to take hold, but by the fear, anxiety, and worry that proliferated around the Klamath case. In this regard, we can see the important role of feelings in generating new worlds, practices, and identities.

For example, many of the irrigators I interviewed acknowledged that avoiding a situation akin to that which unfolded in the Klamath figured heavily into their decision to collaborate with the DRC. While these could be regarded as decisions "rationally" calculated to preserve local livelihoods, in interviews farmers' explanations for choosing to collaborate with the DRC invariably contained emotionally evocative terms, such as "fear" and "worry." Jen, a local farmer, told me that the sole reason that farmers in her area signed on to the DRC's canal piping program was to avoid a Klamath-like event. When asked if farmers chose to pipe in order to maximize hydropower benefits (one of the perks to piping canals is that the resultant pressure can be utilized for hydropower facilities), she responded,

> I don't know. That [hydropower] came after. The [piping] project was sold on retaining our water. We were looking at Klamath Falls and thinking we were going to lose all our water ... We were scared ... There's no way we would have done it if we hadn't seen what happened in Klamath Falls.

Those responsible for garnering support among irrigators for reducing their water use (namely DRC members) described the affective resonance of the Klamath case as a key asset in their endeavors. For example, Matt, restoration manager for the Upper Deschutes Watershed Council (UDWC), meets regularly with farmers to entice them to participate in restoration or leasing projects. Similar to the DRC, the UDWC provides financial incentives to irrigators who help augment flows for fish passage. According to Matt, the Klamath case sparked fear in farmers and helped encourage them to sign onto these conservation projects:

> I can say, hey, you've heard of the Endangered Species Act, you've heard that anadromous fish are coming back into the basin. There's no pressure on you now, but we can help you get a screen or something [so that you're not in trouble in the future]. … And a lot of times that's the way to get our foot in the door. A lot of times they've seen the writing on the wall, they've seen the Klamath and other places, and that's ideally where they start the discussions.

Sean, director of the DRC, said that the threat of becoming "the next Klamath" is what "keeps people at the table" to engage in collaborative practices. And Eric, the director of the UDWC, described the Klamath incident as pivotal in terms of motivating water users to cooperate with environmental institutions. He referred to the Klamath as part of the watershed council's "evolutionary history": "Evolution occurs culturally, linguistically, and all sorts of other ways … People were saying, 'I don't want that [what happened in the Klamath]!' and it became the cultural evolution." According to these water managers, irrigators felt threatened by the potential of a Klamath-like situation unfolding in the Deschutes – enough so that they chose to change their personal water management practices.

Affect, power, and world-making

One afternoon I visited Kate, a progressive organic farmer who inherited her parents' ranch, and helped out weeding the rows of carrots and broccoli. We talked about the Klamath, and the new changes in the Deschutes and why farmers decided to sign onto some of the DRC's initiatives. "Was it really all about seeing what happened in the Klamath?" I persisted. Kate stood up and wiped her hands on her pants. "Does it matter?" she demanded,

> I don't think it matters where the motivation is coming from. I don't know, wouldn't it be wonderful if we were all conservationists and all altruistic and everybody cared as much as me about the fish? … But who gives a shit as long as they're making it better?

In contrast to Kate, I suggest that where motivation comes from does, in fact, matter, and in this regard I point to the importance of feelings and the affective nature of discourse. The Deschutes River Conservancy was born from the aftermath of the Klamath crisis and as such was conditioned by the fear and anxiety that came in the wake of witnessing conflict in a neighboring waterscape. These feelings were channeled into a familiar, albeit misleading narrative – that of "fish versus farmers." As such, the discourse of fish versus farmers carried with it an affective charge, and its connotative power was naturalized as it traveled across sites and as people continued to use it to make sense of their relationships within and to the more-than-human world.

In turning to affect we can better understand how the Klamath case helped inspire a new set of water policies in the Deschutes. For one, affect opens up space for considering how new experiences and new things emerge. While we have certain patterns and habits of response, in every moment we are in a place of transition, and these moments of transition are open-ended; as Massumi (2015, 3) puts it, "[affect] brings a sense of potential to the situation." In the case of the Deschutes, we may consider how the encounter with the Klamath created space for a form of water management to emerge that diverged from the prototypical response to water conflicts we had seen until that point.

Affect also offers us an alternative approach to understanding the operation of power, providing a framework for conceptualizing subjects as produced by discourse as well as by the circulation of emotion and feelings between and within objects and bodies. This move invites us to recognize the force of that which is experientially palpable (while not always clearly defined) and how such structures of feeling work collectively to condition and place limits on possibilities for living and working together. Cultural theorist Ben Anderson (2014, 26) describes the political importance of affect, noting that "States, institutions, and corporations now know, target, and work through affective life." In the case of the Klamath/Deschutes comparison, we cannot help but acknowledge how the "affective symptoms" (McKenzie 2017) of fear, anxiety, and worry were used to mobilize support for particular water policies.

With respect to power, we may also notice that mainstream media, historical accounting, and even contemporary water managers described the events in the Klamath as a crisis. That fear and anxiety were seen to be normative responses to the "Klamath crisis" speak to a particular colonial history. From the perspective of the Klamath tribes, whose adjudicated water rights were recognized, and the short-nosed sucker, the Lost River sucker and the coho salmon, whose rights to regeneration were upheld by federal legislation, the event was perhaps not something to be feared after all.

Ahmed (2004) describes emotions as "sticking" to particular entities, reproducing and maintaining cultural norms. In addition to helping us understand the ways in which feelings matter to water politics and the political potential of such feelings, we also see how certain normative feelings and emotions dominate the public sphere and uphold positions of power. When I searched for any mention of alternative feeling-based responses (such as excitement or relief) to the Klamath "crisis" in mainstream media and scholarly articles, I found them to be invisible. It became clear that fear was the "correct" emotional response to such a set of events. In this regard, flagging moments of hegemonic public affect are one way in which we can reveal the unequal power relations determining seemingly collaborative and equitable water management practices.

Finally, affect can help us to come to new understandings with respect to policy mobility. We live in a world that is invariably entangled – where Klamath sucker fish and BOR projects show up in the Deschutes landscape in unanticipated ways. While geographically distinct, it is clear that narratives, activities, and emotional sentiments traverse watershed boundaries, and in my case disrupted any illusions I may have had about designing an empirical project that could keep the Klamath and Deschutes separate from one another.

I close this section with a comment from Davie, Warm Springs tribal member:

> I was here at that time [the Klamath case] . . . And I think that's the genesis. Water is controversial, and we've always had the rub with irrigation districts because they have the power, and they have the water, and we want the water, and now the stakes are really high . . . and not just in this basin . . . I think nationwide and internationally as well, water has taken on a new meaning for our livelihood here and as we learn more about climate change and about our finite resources and learn about how to prepare for the future, I think that people are having paradigm shifts left and right, and fear causes people to do interesting things.

In observing that "fear causes people to do interesting things," Davie acknowledges the world-making capacity of public feelings and emotions. But his words also remind us that within each felt encounter there is the potential for something new to emerge. As water takes on new affective meanings, even the seemingly intractable legacy of colonial water law (that privileges irrigation over all other uses) can be disrupted.

Conclusion

In the Deschutes and around the world, decisions made around water supply and allocation affect countless lives. It is estimated that in the coming century almost two-thirds of the human population will be living with severe fresh-water shortages, precipitating a range of ecological crises as well as regional and global conflicts. The implications of climate change serve only to heighten these tensions, exacerbating already existing vulnerabilities to water shortages by altering the timing of precipitation and by producing more frequent and extreme droughts (Vynne et al. 2012).

Previous sociological research suggests that one of the major impediments to attaining equitable and cooperative water governance is the failure to recognize the multiple and incommensurable meanings that people make of water and the values assigned to these meanings (Linton 2010; Strang 2004). While limited to a small rural basin in the American West, my study of the Deschutes expands our awareness of human–water relations, demonstrating the limitations of political economic frameworks in helping us understand how and why water management strategies are enacted and upheld.

According to Latour (2004), it is in moments of connection that things are made to matter. We see this in the Deschutes, where the world-making force of feelings helped motivate specific environmental practices from which a different kind of nature emerged; one that is fundamentally relational and historical. This affective approach to water politics is open-ended, and as such it may be that the actions that unfold are not always just or equitable. But by focusing our attention on relationality and change, we may avoid reducing politics to individual self-interests and instrumental values and instead move towards a water politics based on embeddedness, interconnection, and belonging – one that, as Haraway (2008, 62) suggests, encourages us to "engage in a joint dance of being that breeds respect and response in the flesh, in the run, on the course."

Notes

1 Although not the intent of this chapter, it is worth noting that the collaborative management framework has its own set of critiques. By adhering to fixed stakeholder categories, collaborative management paradigms rarely account for the subjective nature of identification or to the dynamic interrelationships between social and ecological worlds. In addition, by prioritizing collaboration as a main goal for environmental politics, collaborative management shies away from conflict and in so doing disables opportunities for meaningful resistance. And finally, collaborative management in many cases excludes people of color, First Nations, and low-income residents from participation (Walker & Hurley 2004).
2 In a recent article, McKenzie (2017) calls for such contributions, proposing that policy mobilities research directly incorporate the role of affect.
3 While those familiar with the complexity of the Klamath case recognize the acquisition of tribal water rights as a key dynamic in the conflict (see Buchanan 2010; Doremus & Tarlock 2008; Gosnell & Kelly 2010), the omission of tribal presence from much of the mainstream media's consolidation of the story alludes to the cultural oppression and silencing of native people that persists to this day.

Bibliography

Ahmed, S. (2004) *The Cultural Politics of Emotion*. New York: Routledge Press.
Anderson, B. (2014) *Encountering Affect: Capacities, Apparatuses, Conditions*. New York: Routledge Press.

Bacher, D. (2009) California Water Wars: Not a Conflict between Fish and People. *AlterNet*. www.alternet.org/story/131157/california_water_wars%3A_not_a_con flict_between_fish_and_people (Accessed March 11, 2010).

Barnard, J. (2013) Report Says Dam Removal Good for Klamath Salmon. *Associated Press*. www.ap.org/en-us/ (Accessed February 12, 2014).

Berlant, L. (2011) *Cruel Optimism*. Durham, NC: Duke University Press.

BOR. (2009) The Klamath Project. *Bureau of Reclamation*. www.usbr.gov/projects/ Project.jsp?proj_Name=KlamathProject (Accessed February 21, 2013).

Buchanan, N. (2010) *Negotiating Nature: Expertise and Environment in the Klamath River Basin*. Berkeley: University of California Press.

Doremus, H. & Tarlock, A. (2003) Fish, Farms, and the Clash of Cultures in the Klamath Basin. *Ecology Law Quarterly* 30(279), 279–350.

Doremus, H. & Tarlock, A. (2008) *Water War in the Klamath Basin: Macho Law, Conflict, and Dirty Politics*. Washington, DC: Island Press.

Fimrite, P. (2013) Fish versus Farmers in Conflict Over the Klamath River. *SF Gate*. www.sfgate.com/science/article/Fish-vs-farmers-in-conflict-over-Klamath-River-4676247.php (Accessed March 13, 2014).

Goodyear, S. (2014) California Fish-Versus-Farmers Debate Rages in John Boehner's Mind. *Water Mark*. https://nextcity.org/daily/entry/california-fish-versus-farmers-debate-rages-in-john-boehners-mind (Accessed January 2, 2015).

Gosnell, H. & Kelly, E. (2010) Peace on the River? Social-Ecological Restoration and Large Dam Removal in the Klamath Basin, USA. *Water Alternatives* 3(2), 361–383.

Haraway, D. (2008) *When Species Meet*. Minnesota: University of Minnesota Press.

Hochschild, A. (2016) *Strangers in Their Own Land: Anger and Mourning on the American Right*. New York: The New Press.

Huitema, D. & Meijerink, S. (2017) The Generation, Diffusion and Impact of Innovations in Global Water Governance. *Journal of the Southwest* 59(1–2), 83–105.

Latour, B. (2004) *Politics of Nature: How to Bring the Sciences into Democracy*. London: Harvard University Press.

Linton, J. (2010) *What Is Water? The History of a Modern Abstraction*. Vancouver: University of British Columbia Press.

McKenzie, M. (2017) Affect Theory and Policy Mobility: Challenges and Possibilities for Critical Policy Research. *Critical Studies in Education*, 58(2), 187–204.

Massumi, B. (2002) *Parables for the Virtual*. Durham, NC: Duke University Press.

Massumi, B. (2015) *The Politics of Affect*. Cambridge: Polity Press.

Miller, J. (2014) California's Drought Is Not About "Fish versus Farmers. *High Country News*. www.hcn.org/ (Accessed March 2014).

Mukhtarov, F. (2014) Rethinking the Travel of Ideas: Policy Translation in the Water Sector. *Policy and Politics*, 42(1), 71–88.

Orr, T. (2014) Why "Fish vs. Farmers" Is a False Dichotomy. *Earth Justice*. https:// earthjustice.org/blog/2014-april/why-fish-vs-farmers-is-a-false-dichotomy (Accessed March 13, 2014).

Overstreet, N. (2014) Water Policy Debate Is Never Just 'Fish versus Farmers'. *Friends of San Francisco Estuary*. http://friendsofsfestuary.weebly.com/blog/ water-policy-debate-in-california-is-never-just-fish-vs-farmers (Accessed January 1, 2015).

Richardson, V. (2009) It's Farmers vs. Fish for California's Water. *Washington Times*. www.washingtontimes.com/news/2009/aug/20/its-farmers-vs-fish-for-california-water/?page=all (Accessed March 20, 2014).

Schutt, R. (2009) *Investigating the Social World*. Newberry Park: Pine Forge Press.

Scott, R. (2010) Coal Heritage/Coal History: Progress, Tourism and Mountaintop Removal. In Gray, H. & Macarena, G. (Eds.), *Sociology of the Trace*. Minneapolis: University of Minnesota Press, pp. 137–167.

Strang, V. (2004) *The Meaning of Water*. Oxford: Berg.

Vynne et al. (2012), August *Oregon's Integrated Water Strategy*. Report for State of Oregon Water Resources Department.

Walker, P. & Hurley, P. (2004) Collaboration Derailed: The Politics of "Community-Based" Resource Management in Nevada County. *Society & Natural Resources* 17, 735–751.

Williams, Raymond. (1978) *Marxism and Literature*. Oxford: Oxford University Press.

Chapter 15

Conclusion

The beginnings of a creative water ethics

Liz Roberts and Katherine Phillips

In our introduction, we meandered through intersecting literatures, providing routes to current thinking about human–water relations, from deep ecology and ecofeminism, to anthropological and urban geographical accounts of modern water systems, from philosophical accounts of the more-than-human and bioethical, to accounts that seek to bring spiritual, poetic, sensory and aesthetic dimensions creatively to life. These influences are found across this collection. With a specific focus on creative approaches and arts practices, the authors draw out the potential of creativity to inform these different literatures. As greater than the sum of its parts, we wish to elaborate on the central contributions of the collection as a whole, thinking about how it speaks back to current water research and to point towards future avenues of research and collaboration. Specifically, we hope that this work contributes towards the development of a creative water ethics.

We begin by articulating how the the collection responds to a call for starting from a fundamentally different analytical (and even, subject) position in order to do research on human–water relations. The contributors forward alternate ways of knowing through changing the subject/object relationship of knowledge-making, incorporating 'others' into it via creative processes and by using arts and creativity to make visible alternate relations with water. Second, we consider how the volume represents an attempt to bring other 'voices' to the fore that would not ordinarily be a part of water resources policy or management, through adopting a number of creative 'tactics' and relations with watery others. We then illustrate how 'agency' has been central to many of the chapters and how creative approaches can expand the meanings of agency in productive ways. Finally, we rehearse some potential pitfalls of creative approaches and share learnings from practitioners and academics in this collection, also thinking about where these experimental 'tributaries' might meander next.

Process, plurality and making visible alternative spaces

The first thing that is clear is that projects with creative and participatory elements are helping to foster different human–water relationships through incorporating lay/local/plural knowledge(s) into water governance, and through cultivating an 'ethic of care' by paying special attention to watery places, practices and habitats. They draw attention to encounters with water outside of the hydrosocial contract which creates identities of 'water providers' and 'water consumers' and outside modern water's instrumental language of 'resources', 'systems' and 'services'. Exchanges happen in these different types of encounters with water that help to reinforce new sensibilities, and a 'trace' of such encounters filters into everyday practice as a latent form of knowledge that can be drawn on. Creative practices can enable encounters that can be effective in reflection and learning processes, having the capacity to 'spur ideational change and those who have the capability to invoke that change' as suggested by Farnum et al. in Chapter 8. Creative methods can also consolidate new ideas or sensibilities, born out of social exchange, and communicate them to a wider sphere, drawing creativity out of others and 'giving voice' to them, as illustrated for example by Leeson in relation to the Geezers on the Thames (Chapter 1), and Bakewell et al. in relation to collaborative water governance (Chapter 4). Critically, the chapters respond to calls for alternate spaces, models and narratives for human–water relationships.

In looking at the collection we are able to draw out connections between the chapters that together contribute to a wider narrative. Farnum et al. call for studies to be conducted with people who view water in fundamentally different ways (Chapter 8), resonating with Rudestam's argument that equitable and cooperative water governance is failing because of a lack of recognition of 'the multiple and incommensurate meanings that people make of water and the values assigned to those meanings' (Chapter 14). The chapters, as a collection, suggest that formal/informal binaries of water and of water governance, as Big/modern water versus experiences outside of that, are not so straightforward, and perhaps these sit alongside each other in how people actually experience water in their everyday lives. Creative methods can make visible alternative practices and encourage reflection over alternative spaces of encounter and how these interact with techo-managerial spaces. For example, *La Rasgioni* performance – a type of theatrical community meeting to make local decisions originating from Sardinia – provides a more informal and communal way of discussing environmental issues (Bakewell et al., Chapter 4). Likewise, a community art project explores how the power of the river Thames can be used to support local communities by seeking to 'create alternative models and demonstrate their effect' (Leeson, Chapter 1). The emotional spaces of water are also shown as a suitable alternative site of inquiry, illustrating the political force of feelings that reinforce particular water-place narratives as central to understanding how and why water policies are adopted (Rudestam, Chapter 14). In different ways, these are all engaging with the political, whether this be through community water conflict resolution, political acts of 'making visible', or through examining intangible aspects of 'rational' water policy discourse. Such approaches help us see how world-making happens through the 'speculative, imaginative and engaging forms of politics propagated by creative practices' (Kanngieser 2013, in Hawkins 2018, 20). Meisch (Chapter 10) warns that arts and humanities scholars should be sceptical about different forms of knowledge being subsumed within techno-scientific rationales that seek to abstract and generalise, yet such approaches do help to re-frame techno-scientific policy positions as a problem, rather than the position from which to be offering solutions.

One reason for pitting such positions/disciplinary perspectives as the problem is their tendency to be reductive, simplistic and homogenising, as argued by Hoolahan and Browne (Chapter 12), and Meisch (Chapter 10), in this volume (see also Strang 2016). The collection reinforces the plurality and significance of personal and cultural meanings, and values associated with water that are far from simple, seeking to examine their complexity and not necessarily try to resolve it. Linked to this celebration of pluralism, is a commitment to recognition of the open-endness of human–water relationships, in terms of creative processes, knowledge creation and decision-making. A processual or open-ended understanding of meaning, knowledge and being means that the creation of lifeworlds becomes a type of ethical relation with an other(s).

Part of shifting emphasis away from creator and creation (an end product) to something more process-focused and open-ended is an acknowledgement of the multiple actors and 'actants' that help to shape meaning. In many of our chapters, agency is given over to audiences (readers, listeners) and co-producers (communities, organisms, rivers) as a distributed creativity. For example, Gorell Barnes (Chapter 2) describes her decision to leave her writings and map-making as an assemblage of her experience, choosing not to write over her art-practice with a cohering narrative that fixes things. Instead she allows her struggles to make sense of her collected memories and materials to be visible, all the while acknowledging that the reader(s) will take these forward and shift the meaning, bringing to bear their own experience through their interpretations. Likewise, Lyons (Chapter 3) claims that '[t]here is no single thread nor argument in the streams of watery activations and flights of fancy described here' to describe his deep mapping as an assemblage approach, while Meisch (Chapter 10) argues that it is reductive to say what a creative form like historical hymns offer policy frameworks as different readers and audiences from different historical and cultural contexts will take different things from it. For Hartley (Chapter 7), more attention needs to be given to the context of production (of knowledge, of cultural forms) 'given the distributed, multi-scale nature of change in the Anthropocene', and St John (Chapter 9) calls for researchers to 'pay attention to the way life is bought into perceptual being'. Their process-focussed approach suggests that knowledge about human–water relations is never complete, and to universalise and fix it through scientific or instrumentalist language disallows other connections and relations to be made or other voices and forms of agency to be elevated. Through our broad conceptualisation of creativity we also wish to detach discussions of 'meaning' from fixed representations and outputs or official forms of knowledge. Through the chapters' explorations of artistic processes and embodied practices, meanings are relational, emergent and changing, captured fleetingly in ways we might not expect, such as through the playing back of an audio tape loop degraded in river water (St John, Chapter 9) or the expression of a surfed wave experienced sensorily and lost immediately (Anderson and Stoodley, Chapter 6).

De-privileging anthropocentric and dominant accounts

The ethos behind many of the chapters chimes with broader debates around human–environmental relations, especially how they are theorised within the academy. Current water policies and management strategies continue to face critique as being underpinned by conceptual assumptions about nature and culture as separate domains. Strang (2016) argues that the idea of nature as 'other' permeates every form of engagement with the non-human, including water policy. A drive to change this view can be seen from diverse literatures, and is reflected in a focus on indigenous knowledge and practices as well 'multi-species' enquiry, which seeks to give less anthropocentric accounts of human–environmental relations by highlighting that the ways that human, non-humans and even technologies interact is the result of dynamic processes (Strang 2016). Much of this work is inspired, as noted in our introduction, by Haraway and colleagues' development of 'interspecies ethnography' (Haraway 2008; Kirksey and Helmreich 2010), which put the non-human in the position of the subaltern and seek to give them voice by adopting a non-human standpoint (Strang 2016). There is a growing body of work that seeks to 'give voice' to the non-human through experimentation and creative approaches such as those in our chapters have much to offer.

In an effort to de-privilege anthropocentric accounts and meaning regimes that support the power relationships inherent in modern water infrastructures, particular 'tactics' or practices are promoted by our contributors. This involves an ethical or creative 'attunement' to the often invisible co-producers of human–water relations (including animals, organisms, habitats). In our chapters, this is variously referred to as 'conscious reading' (Meisch, Chapter 10), an 'ecological sensibility' (Hoolahan and Browne, Chapter 12), 'attentive listening' (St John, Chapter 9), an 'unconscious optics' or 'psycho-poetic intuition' (Lyons, Chapter 3) and 'story-listening' (Bakewell et al., Chapter 4). These are efforts to bring about an 'intensifying of our perceptive abilities' (Meisch,

Chapter 10), to be able to compare different perspectives, to cultivate empathy through the creation of life-worlds (Bakewell et al., Chapter 4; Foley, Chapter 5; Meisch, Chapter 10), and heighten the role of emotions and feelings in generating new worlds (Anderson and Stoodley, Chapter 6; Rudestam, Chapter 14). Narratives and storytelling are shown to be an important aspect in these creative practices. They help to reconnect water habitats and potable water supply after the 'experiential distancing' that happens in Modern water systems (Lyons, Chapter 3); sometimes this is understood by artists as their ecological role or imperative (Gablik 1991). Our chapters seek to privilege other (human) voices previously excluded from water governance (e.g. Gorell Barnes, Chapter 2; Birkinshaw, Chapter 11) and other species and forms of agency (e.g. St John, Chapter 9; Rudestam, Chapter 14). St John points out that creative practices are foregrounded in multi-species enquiry as a means of troubling human/non-human boundaries. Indeed, creative processes can help us feel the 'liveliness' hidden in things and 'reveal threads connecting their fate to ours' (St John, Chapter 9). Lyons identifies creativity as taking on a new role in public discourse due to anthropocenic change and shifting social-ecological relationships. The chapters illustrate that there is a creative ethic that can inspire more equal human–water relations.

Articulations of 'agency'

A third contribution of the collection can be seen as an exploration of the notion of 'agency'. Many of the chapters seek to elevate the agency of water in various ways as something that has the capacity to act on (humans) and to contribute to meanings associated with it (e.g. Hartley, Chapter 7; St. John, Chapter 9). Some of the authors explicitly draw on literatures on materiality and Actor-Network Theory (see Bennett 2009; Latour 2009), while others arrive at the idea of the agency of water in other ways. Agency is distributed, found in relations between things, rather than purely a characteristic of humans. The material affordances of water, watery things and species co-constitute their meaning as they come into relation, alongside cultural and symbolic contextual affordances. While staying attentive to the often destructive power imbalance that humans, for the most part, uphold over nature, our authors explore this type of material agency, in varying attempts to rework the relationship between humankind and 'the other'. They adopt a (micro-) political positioning or ethical imperative that puts 'non-human agency at the fore' to challenge forms of water resource exploitation (Alberti 2014, 160; Strang 2014). It is this type of ethical imperative that we take as the basis of creative interventions presented in the chapters, with many contributors responding directly to these or parallel human–environment debates.

Yet to view the forms of agency that the authors suggest within human–water relationships in strictly these terms would impose a limitation, and would not grant the opportunity to explore more of the distinct contributions that creative approaches give to this topic. The chapters draw on the concept of agency: to simply challenge dominating power relationships; to refer to something akin to having a respect or respectful relationship with water and its shared spaces and inhabitants; to variously mean the capacity of water, of people, and of organisms, habitats and materials to act upon something, as a sense of potentiality or an affordance; to describe an elemental or affective force that holds water in a creative or embodied 'pull'; to a disruptive force or 'encounter' that forces us to think; and finally, in terms of connectivity. As a central motif, we'll take a moment to expand on these.

Water can have materially disruptive agency when it effects people's lives through appearing in quantities that are more or less than the usual, expected amount. In Birkinshaw's chapter (Chapter 11) it is the material qualities of water that disrupted the political economy of water supply on the edge of Delhi and created precarious new constellations of power. But water can also disrupt through the meanings that get attached to it, such as the way that the Klamath region's 'farmers vs fish' narrative erupts in the Deschute area of North America in a way that implicates decisions made about water policy despite the unlikeliness of similar impacts in such different catchments (Rudestam, Chapter 14). Water, no matter how much we seek to control it, will always retain something wild about it (Edgeworth 2011), leaking, seeping or rupturing out of containment. In these chapters and others (Anderson and Stoodley, Chapter 6; Hartley, Chapter 7) the material and cultural affordances of water (such as a wave for swimming and surfing in the form of an affective pull and iterations of identity) can be seen as ways in which water itself influences the meanings that humans associate with it.

Some of the authors also attribute a type of elemental agency to water that inspires, is given attention or bubbles up in their creative practice. Language used to describe water often involves almost magical or spiritual terms, talking about its 'pull', its 'draw', its 'power'; the coast becomes 'alchemical' or 'magnetic'. In Foley's chapter (Chapter 5), one swimmer compares getting into the water to the part in *The Wizard of Oz* film where it transforms from black and white film into colour. This metaphor describes a sentiment that many people feel towards watery activities and landscapes. Elsewhere, the long-term relationship between special meanings and water sources or confluences have been noted (Edgeworth 2011). Other chapters describe the therapeutic effects and relationships with belonging that water can give (Gorell Barnes, Chapter 2; Anderson and Stoodley, Chapter 6). Leeson (Chapter 1) describes how her work has unintentionally returned over and over again to the River Thames because of its historical, symbolic and transformative power, while Lyons (Chapter 3) argues that humans have forgotten these mythic, symbolic, magical and subconscious aspects of water. The chapters illustrate how creative approaches can help articulate these special and sometimes intangible relationships.

Part of the strength of drawing on creative and arts-based engagements with water is that they may help to – following Bennet (2009) – 're-enchant' water and illustrate a distinct, attentive relationship or attunement with water materially, as agential and a vital or energising matter. In techno-scientific derived disciplines there is a wariness to engage with the idea of giving water too much agency, with a fear that it errs towards ideas of sentience and animism. Strang, herself, notes that '[i]t is important not to assume some form of intentionality or sentience or to 'fetischize' material objects' (2014, 139). In creative projects it is this type of engagement that may be most powerful. Certainly, historical, anthropological art-forms and religious art has a close relationship with animism (in terms of iconography and symbolism) and indigenous cultures have different relationships with water via deep attachments to place, totems and spiritual objects. Nature has equally been attributed 'subtle metaphysical qualities' in Western romanticism and nature writing (Lyons, Chapter 3). Disciplinary expectations permit more freedom within creative projects and the arts and humanities to explore alternative forms of agency that do not fit within particular types of scientific language or rationales.

The idea of non-human energy is more common in cultures that do not privilege techno-scientific modernist frameworks in the same way. This can be thought as an animated perception inherent in nature connections. A form of subtle energy can be found in *qi/ki* in China and Japan, as *prama* in India and as *atua* in Maori (Flowers et al. 2014) alongside many more examples, however:

> [t]he lack of Western academic consensus regarding its very existence, and the challenge of finding a language to describe it, relegates the knowledge gained from using modalities that profess to work with subtle energies as naïve, impossible, and often, inconsequential.
>
> (Flowers et al. 2014, 113)

Through different methods, many of the chapters highlight the 'subtle energies' of different forms of water, such as waves and rivers. An attention to subtle energies and modes of enchantment does not automatically leap toward material determinism. Across the collection, we can clearly see the subtle political and cultural affective energies of water-related issues. We propose subtle energies might be further brought into water research and celebrated as a mode of knowing.

Another form of creative agency found in our collection is through the idea of connectivity. Echoing Strang's assertion that it is a relational agency that can be found in the material qualities of water, many of our authors take inspiration from ANT and theories of affect to consider the way that humans come into constantly changing constellations or assemblages with other 'actants' including their environment, which co-constitutes both their experience and the meanings attached to it. The way water moves and the forms it takes has inspired our contributors to use it to describe how knowledge is created (Hartley, Chapter 7; Bønnelykke, Chapter 13) through such assemblages. The wave functions as a metaphor where form is always tentative, coming undone and re-forming into new patterns through flux. This stands for the way that individuals make connections and conduct constant tactical improvisations and experiments in their everyday lives that rework connections and create new relationships.

Creative approaches can bring attention to the creative potential in everyday moments in a more overt way. In our introduction, the improvisational aspects or tactics for creativity were highlighted, where creativity is not cut off from mundane and everyday cognition and practices. We cited Hallam and Ingold (2007) who challenge the widespread understanding of creativity as 'the new', as innovative and exceptional, standing out from what came before as radically different. Instead, emphasising a forward-looking creativity, which is improvisational and relational and where life is an ongoing series of improvisational and creative tactics as people and objects bump into each other in different environments, opening myriad possibilities for relations. Creative practice positions things in 'generative juxtapositions' (St John, Chapter 9). Through thinking about connections, a 'contingency awareness' (Meisch, Chapter 10) can be cultivated as a type of ethic or empathy: a disposition to recognise alternatives. As such, an assemblage approach is taken by several of our authors, which allows for this creative connectivity to be plural and open-ended (e.g. Gorell Barnes, Chapter 2; Lyons, Chapter 3; Foley, Chapter 5), as a deliberate creative method. Creativity is also found in our chapters to be processual and emergent out of the everyday, involving a re-making and transformation of social practices in everyday life. Understanding creativity as both a professional skill and as informal, vernacular and amateur is helpful in this context, avoiding the policing of what can count as 'creative', and as offering something 'differently valuable' (Hawkins 2018).

Within this type of everyday creativity is the potential for a radical or transformative 'encounter'. Several of our chapters discuss the relational agency of water as an 'encounter' (Anderson and Stoodley, Chapter 6; St John, Chapter 9; Hoolohan and Browne, Chapter 12). An encounter is theorised as a pause or reflective moment that is caused by a rupture in habitual ways of thinking or being. This might be the result of an affective force or might occur out of repetition, when the same becomes dissimilar; in the example of the wave, new patterns are formed out of the old, and a reflection or reconfiguration of our understanding is needed. When we think of creativity as related to everyday practices, a swim, a surf, a river clean-up or a 'way of life' for islanders at risk from rising sea levels might create difference through repetition via regular engagements with water. In this way, this type of creativity becomes a type of micro-politics, in the form of a resistance or a localised change, or through embodied types of knowledge, which becomes the source for a more overt politics (see Anderson and Stoodley, Chapter 6, where surfers become environmental advocates for their local surf spots). It can also be seen via an ethic of care for the ocean or rivers and the organisms that live in it as cultivated through a close relationship with them (Foley, Chapter 5; Hoolohan and Browne, Chapter 12). As Leeson points out in Chapter 1, 'Change (where one is)' is a form of political power. Our chapters show that watery identities emerge out of 'encounters' with water, through practices and through narratives of place that inspire a particular type of relational ethic. They highlight the different sites and scales at which human–water relations can be understood: at the scale of the body, the community, the micro-organism, through narrative exchanges and within creative processes.

Water can function as both the material and environment that makes creativity possible. Through creative practices this can be made explicit and scaled up to affect wider audiences. An encounter can be manipulated through creative methods, which are sometimes also viewed as ethical, to 'render things strange' (St John, Chapter 9), or to focus on the 'hidden details of familiar objects' (Lyons, Chapter 3). Creativity can open new spaces of encounter (Bennet 2009). This type of creative change may be one potential transition out of the lock-in of current socio-technological systems that comprise 'Big' or 'modern' water and the hydro-social contract, as a form of change that is iterative and starts small-scale. For example, none of our chapters frame the individual as a consumer, and none simplify human relationships with water as access to potable supply, or use the language of 'resources' or 'services'. Nevertheless, as Lyons says, the 'powerful forces of status quo' should not be underestimated (Chapter 3). Hoolohan and Browne (Chpater 12) describe how our expectation for an endless supply of clean water is 'baked in' to everyday routines and practices in Western water infrastructure. They critique the current framing of sustainability interventions, as working within a paradigm that views 'modern' water systems as the norm and the only possibility within a neoliberal context. Instead they call for policy options that move away from placing responsibility on the individual water consumer and towards a more holistic approach for water conservation that recognises the interrelationships between humans, water, animals and habitats.

Dangerous neoliberal waters?

So far in this conclusion, we have shown the contributions that creative approaches can make to understanding or reframing human–water relationships. It is worth also considering some of the dangers that might surface when creative methods or practitioners are enrolled in inter- and trans-disciplinary academic projects. We have seen this approach increase across social science disciplines such as archaeology, anthropology, sociology and geography, with an interest in creative practitioners as more than external figures of interest, but with creative practices as part of the 'doing' of knowledge-making (Cochrane and Russell 2014; Hawkins 2014, 2018; Morgan 2009). This 'creative turn' is more of a (re)turn where there are important lessons from previous forays with creativity. For example, early geographers and explorers, anthropologists and scientists were keen to place art at the heart of scientific development and as a way of engaging the public, through conveying the 'geopoetics' or aesthetics of their discoveries (Hawkins 2018). We can also look to the way that meanings were previously constructed around nature in paintings. Art history shows that cultural tastes dictated that nature often be viewed as part of the rural idyll or 'wilderness', symbolically recreating particular power relations, often with humans and exploitative activities omitted, giving a sanitising and othering effect to nature. This effect has been repeated more recently in the creative economies and creative cities agendas where a colonising of artistic practice has had a sanitising effect on city centres and previously culturally and socially diverse neighbourhoods through regeneration projects. Across these, process is less important than final representations.

It is worth looking back to this history and having an eye open to the types of appropriation that can occur, especially in light of current neoliberal contexts within the University. Tolia-Kelly (2011, 137) notes that 'university funders are bounding towards a culture of impact and public engagement' enrolling visual culture and arts along the way. There is the perception with this type of work that it is 'interdisciplinary, forward thinking and relentlessly positive' (Hawkins 2018, 13). We should also acknowledge the problems that can be attached to such arty engagements, especially if they are 'parachuted in'. There is a risk that such projects seek an artistic output that can be used to engage the public and over look other important aspects of creative process as a result. The creative practitioner becomes viewed as a translator of research already done, rather than a facilitator or active agent within the research process. Equally, creative practitioners can become part of the workforce that universities 'extract labour from without appropriate value structures' (Mclean 2017) and their engaged practices might be a 'slow' form of knowledge creation that doesn't quite fit with the 'fast' academy in terms of funding timeframes (Hawkins 2018).

Our contributors shared stories of their own experiences. One academic–arts partnership was unable to apply for a recent funding call with a creative focus due to the research body stipulating that artist salaries were not incorporated as part of the grant. Working within multidisciplinary settings with other sectors requires greater appreciation of external structures, such as the understanding that an artist would not automatically have alternative sources of income. Another contributor commented that it was difficult to fit within the timeframes of 'fast' academic projects and it could work counter to her own approach to take time to build a portfolio of people, groups and communities who offered financial security through ongoing work together. These relationships and related security could be jeopardised by academics external to this who, working via the artist, seek 'quick wins' and outputs but are insensitive to context and dimensions of trust. Individuals seeking to do arts practice research within social sciences departments also find themselves falling between the cracks of existing university structures such as assessment protocols, and one contributor noted PhD work that lost the 'richness of the art' in order to 'fit' (also noted by Hawkins 2018). A final contributor commented that storytelling and creative approaches could be misused when they are applied in a uni-directional or functionalist way, such as the case with boring stories with a too obvious moral punchline. While the arts can sometimes play with manipulation and this can be viewed with skepticism, we could also invite this ambiguity and give credit to audiences, rather than having creative approaches function simply to 'colour the pictures of a preset scientific or economic message'.

There are also political opportunities within creative approaches, as we have illustrated in our chapters, through the blurring of boundaries between human and non-human as a form of water ethics, and through art as a 'politics in action'. Yet there is the danger that it becomes a fad, without critical reflection on the processes and politics involved, and as a form of disciplinary colonialism. It has been proposed that we need to '[r]emain sufficiently vigilant and critically aware to ensure they do not become a parody of themselves, something wholly corruptible and able to be put to use in exactly the opposite ways as those for which they were intended' (De Leeuw et al. 2017, 6).

As Hawkins (2018, 22) adds, it is important to 'temper our excitement over the political opportunities of particular modes of creativity within research projects with a careful reflection on the politics of our own practices'. This is a concern for both artists working with universities in interdisciplinary projects, and artists working independently but with neoliberal partners and in other environments. What many of the chapters cleverly show is that artists and creative projects can contribute to knowledge about their environment without adopting a subject position outside of that. The types of creative methods and 'hybrid ontologies' (St John, Chapter 9) that appear in social sciences and arts and humanities seek this same 'being with' position as a basis for experience, as opposed to an object/subject position; this is a position that can never be adopted by academics or policy makers seeking economic or technical rationales for framing relationships with water because you cannot externalise costs of human exploitation if you acknowledge that we share the same relational web.

Through efforts to become more interdisciplinary, more participatory or accessible to the public, arts and creative methods are being adopted in global challenges and wicked problem research, including issues around water. So far efforts to increase the scope of water policy research has been limited and remain within a techno-managerial framework. These efforts have been described as 'half-hearted' with the most common approach seeking to inject social data such as 'key variables' of human behaviour into analytical and agent-based models used in the natural sciences (Strang 2016). Recent work in 'social hydrology' can be more or less reductive in this way, often limiting real exchange of knowledge; in light of this, Strang (2016, 25) calls for 'less compressive methodologies: ways of bringing different datasets into conjunction without condensing their meaning'. Creative approaches are one way to respond to that call, drawing out and questioning different and plural meanings and values tied up in human–water relationships, and also allowing them to be 'affective' as a transformative 'pause' or 'encounter'. As a collection, we have sought to illustrate the ways that individuals and communities can participate in and frame understandings of relationships with water and environment that can provide a basis for changed practices.

One of the major contributions of this book is the bringing together of contributions that illustrate through their creative engagements the material affordances and creative potential of water enabled and enacted through 'everyday' human–water interactions. These are reflected upon in detail, where, in contrast, previous volumes may have taken such interactions as mundane, normative, taken for granted, subconscious, unimportant, apolitical. Yet, alternative knowledges might enable different sensibilities to be fostered through more meaningful and reflective watery relationships, in response to Krausse and Strang's (2016) argument to cultivate a 'water ethic'. The chapters provide avenues away from more scientific and technical literatures that position the 'knower' as being on the outside of their environment and create a rationalising distance between them. Meaning, representations and knowledge are all closely linked with power as a way of producing, reproducing and maintaining power relations. Through making visible and giving significance to alternatives, we can begin to shift power relationships and chip away at the dominance of modern water systems and associated discourse and ideology. A creative 'water ethic' might allow for alternate patterns of use and management. The human–water relationships given space in the book evidence a different understanding of creativity found in the micro-politics of everyday embodied improvisations, iterations and tactical adaptations, illuminating different aesthetics and values associated with water. We argue that within these lies the potential for something more explicitly and politically transformative to be elicited.

References

Alberti, B. (2014) How Does Water Mean? *Archeological Dialogues*, 21(2), 133–150.
Bennett, J. (2009) *Vibrant Matter: A Political Ecology of Things*. Durham, NC: Duke University Press.

Cochrane, A. and Russell, I.A. (eds) (2014) *Art and Archaeology: Collaborations, Conversations, Criticisms*. One World Archaeology Series, Volume 11. New York: Springer-Kluwer.

De Leeuw, S., Parkes, M.W., Morgan, V.S., Christensen, J., Lindsay, N., Mitchell-Foster, K. and Jozkow, J.R. (2017) Going Unscripted: A Call to Critically Engage Storytelling Methods and Methodologies in Geography and the Medical-Health Sciences. *Canadian Geographer*, 61(2), 152–164.

Edgeworth, M. (2011) *Fluid Pasts: Archeology of Flow*. London: Bloomsbury.

Flowers, M., Lipsett, L. and Barrett, M.J. (2014) Animism, Creativity, and a Tree: Shifting into Nature Connection Through Attention to Subtle Energies and Contemplative Art Practice. *Canadian Journal of Environmental Education*, 19, 111–126.

Gablik, S. (1991) *The Reenchantment of Art*. London: Thames & Hudson.

Haraway, D.J. (2008) *When Species Meet*. Minneapolis: University of Minnesota Press.

Hawkins, H. (2014) *For Creative Geographies: Geography, Visual Arts and the Making of Worlds*. New York: Routledge.

Hawkins, H. (2018) Geography's Creative (Re)turn: Towards a Critical Framework. *Progress in Human Geography*. Available on Online First.

Hallam, E. and Ingold, T. (2007) *Creativity and Cultural Improvisation*. London: Bloomsbury.

Kanngieser, A. (2013) *Experimental Politics and the Making of Worlds*. London: Routledge.

Kirksey, S. and Helmreich, S. (2010) The Emergence of Multispecies Ethnography. *Cultural Anthropology*, 25(4), 545–576.

Krause, F. and Strang, V. (2016) Thinking Relationships Through Water. *Society and Natural Resources: An International Journal*, 29(6), 633–638.

Latour, B. 2009. *Politics of Nature*. Cambridge: Harvard University Press.

McLean, H. (2017) Hos in the Garden: Stating and Resisting Neoliberal Creativity. *Environment and Planning D: Society and Space*, 35, 38–56.

Morgan, C. 2009. (Re)Building Catalhoyük: Changing Virtual Reality in Archaeology. *Archaeologies*, 5(3), 468–487.

Strang, V. (2014) Fluid Consistencies. Material Relationality in Human Engagements with Water. *Archeological Dialogues* 21(2), 133–150.

Strang, V. (2016) Re-Imagined Communities: A New Ethical Approach to Water Policy. In K. Conca and E. Weinthal (eds.), *The Oxford Handbook of Water Politics and Policy*. Oxford; New York: Oxford University Press, pp. 142–166.

Tolia-Kelly D.P. (2011) The Geographies of Cultural Geography II: Visual Culture. *Progress in Human Geography*, 36, 135–142.

Afterword

Interview with Matthew Gandy, Professor of Cultural and Historical Geography, University of Cambridge

We have aimed with this volume to bring together a diverse range of voices and approaches focused on human–water relationships. As with any good interdisciplinary effort, it is necessary to appreciate the full extent of contributions made from within – as well as across – disciplinary communities. One of the areas of research and writing into human-water relationships that has in recent years offered particularly interesting insights into human–water relations is that of urban political ecology. We asked Matthew Gandy, author of *The Fabric of Space: Water, Modernity and the Urban Imagination* (2014) to share some thoughts on where he feels water research is going and what role creative and imaginative approaches have in developing our methods and understanding.

1   *What from your perspective is the novelty in urban political ecology research on water at the moment? What are the growing trends you anticipate?*
    I think that water has played a prominent role in urban political ecology research since its inception in the 1990s. There are still many interesting books and papers being published on the theme of water from a broadly urban political ecology perspective but there are several trends we can discern. I think the relationship between water and state formation has become a more prominent focus with a particular emphasis on the construction of major engineering projects such as dams, drainage schemes and complex water supply systems (examples include the work of Erik Swyngedouw on modern Spain). There are also parallels between the emphasis of urban political ecology on water, power and space and insights from environmental history (examples include the work of Sara Pritchard on France and David Blackbourn on Germany). Another strand of work, especially associated with anthropology, has been an interest in water and everyday life, using a variety of ethnographic insights. The recent work of Nikhil Anand on Mumbai really stands out with his subtle reworking of the connections between infrastructure, modernity and different understandings of the public realm in a global South context. Another key work is Antina von Schnitzler's analysis of post-Apartheid

infrastructure politics and the role of infrastructure systems in securing political legitimacy. A further area of emerging interest is the connection between water, epidemiology and corporeal geographies of human health, a nexus of socio-ecological relationships that now has added poignancy because of climate change. It would be great to see UPE perspectives applied to the Zika virus, for example, perhaps linked with epigenetic research into insect behaviour. The 'ecology' dimension to UPE should be taken seriously as a dynamic scientific field in its own right rather than a merely rhetorical accoutrement to conceptual and political frameworks.

2   *How would you interpret 'creativity'?*
The term creativity can be approached from a variety of angles. The idea of creativity relates to the research imagination and the possibility to conceptualise socio-ecological dynamics in novel ways. In this sense I think comparative research can be helpful in generating unexpected juxtapositions and blurring unhelpful typologies. Urban infrastructures clearly unsettle existing categories and are suited to interdisciplinary research methodologies. From an architectonic perspective, of course, it is necessary to imagine alternative spatial forms as a precursor to potentially bringing these new structures into being: the spaces of the real, the virtual and the imaginary are often combined. Creativity is also present in different modes of writing or representation: the precise choice of words or images, their sequence and arrangement and the precise contexts in which they are experienced and communicated.

3   *What role has creativity played in your own research?*
Cultural representations of water, water infrastructure systems and different aspects of everyday life have always fascinated me and often served as a creative starting point for wider reflections on nature, modernity and urban space. I remember, for instance, first encountering the photographs of the Paris sewers, taken by Félix Nadar in the early 1860s, and being intrigued by this orderly and spacious underground realm. Using a range of historical sources I began to explore the complexities of Second Empire Paris as a highly ambiguous stepping stone towards the technological modernism of the twentieth century and its associated forms of networked space. More recently, I have been intrigued by the future scenarios for London developed by the digital design studio Squint/Opera: these representations of a flooded London in the year 2090 play on the creative re-imaginings of a near future: they provide an alternative iconography to both dystopian science fiction and technocratic forms of engineered resilience. In terms of visual resources I have also been interested in cinematic responses to urban space and I have used film making as a specific kind of research methodology in Mumbai, and more recently in Berlin.

4   *What contribution do you think creative approaches could have to urban political ecology research on water and broader watery research?*
I think creative approaches widen the scope for interdisciplinary research: there is an openness to different kinds of data or interpretative frameworks. To accept that urban political ecology is not the domain of a small set of disciplines is significant but we are left with the need to develop conceptual vocabularies that can operate effectively across disciplinary boundaries. It is striking that 'watery research' is sometimes confined to a narrow conceptual frame because the object of enquiry is erroneously assumed to be clearly defined. The ubiquity of water can paradoxically work against the development of more conceptually innovative frameworks for analysis or interpretation.

5   *How might this speak back to conceptions of 'modern' water?*
One interesting example is the photographic representation of water towers by Bernd and Hilla Becher. These taxonomic sequences of endlessly varied architectural forms can be read as a eulogy to the industrial landscapes of Europe and North America but also a nod towards the taxonomic order of modernity with these curiously shaped buildings resembling the serried ranks of life forms to be found in a natural history identification guide. The curatorial aesthetic adopted by the Bechers for these images produces a kind of ambiguity between past and present: we cannot be sure whether these structures are still in use or merely relics in the landscape. Intersections between infrastructure and landscape produce an endlessly fascinating series of spaces that can transcend their narrowly functional aesthetic to become integral features of modern culture. Another poignant example is the 51-mile concrete channel of the Los Angeles river: this mundane yet extraordinary infrastructure system has recently become a focal point for multiple re-imaginings of relations between nature and culture in the LA metropolitan region, ranging from crudely nativist approaches to ecological restoration to intricate forms of cosmopolitan ecology. Less clear, however, in these emerging ecological imaginaries for Los Angeles is the place for collective memory and various celebrations of these concrete landscapes in popular culture.

6   *What got you interested in doing water research?*
I think I initially approached water through the lens of urban infrastructure and urban metabolism: my PhD was focused on recycling and waste management but many of the themes clearly touched on other dimensions to urban space such as energy, water and environmental politics. I also remember reading a wonderful PhD thesis on the history of London's water supply by Asok Kumar Mukhopadhyay that explored the intersections between politics, engineering and urban space. My first significant work on water was on the construction of the New York water supply system and the

emergence of an 'ecological frontier' for the metropolis located deep in the Catskill mountains. Subsequently I've developed water-related research in different directions including the development of infrastructure systems in the global South, corporeal dimensions to modernity and connections between water, climate change and bio-diversity.

7 *Do you have a film, photograph, installation, song with water as its theme that you particularly relate to, or are inspired by?*
There are many examples of cultural responses to water, or cultural depictions of water, that have influenced my work. A few examples would include: Émile Zola's novel *L'Assommoir* (1877) and its evocation of everyday encounters with water such as communal laundries before the era of widespread access to modern plumbing; Fela Kuti's song 'Water No Get Enemy' from the album *Expensive Shit*, first released in 1975, that I heard while doing fieldwork in Lagos; and Andrea Arnold's waterside interlude in her film *Fish Tank* (2009) that is set on the eastern edge of London.

8 *What contribution do you feel this collection makes to water research?*
Beyond the empirical richness of the collection the most striking themes for me that emerge from the chapters are a range of conceptual explorations at the leading edge of water research. I was very interested to see emerging interest in different forms of 'attunement', attentiveness and materiality, posing questions in terms of research methodology as well as the interpretation of different kinds of developments that span human and other-than-human realms. The looming presence of the Anthropocene can also be felt in some of the contributions: how, for example, does an acknowledgment of 'deep time' in terms of the history of human relations with water relate to contemporary developments, and a sense of compression or acceleration across a range of fields? How can we reconcile calls for 'slow' thinking, with room for reflection, advanced by Harriet Hawkins and others, with the pressing sense of urgency posed by environmental challenges and also the increasingly neo-liberal context of the academy? Where can we find space and time for creative thought?

Index